Logic, Epistemology, and the Unity of Science

Volume 44

Logic, Epistemology, and the Unity of Science aims to reconsider the question of the unity of science in light of recent developments in logic. At present, no single logical, semantical or methodological framework dominates the philosophy of science. However, the editors of this series believe that formal techniques like, for example, independence friendly logic, dialogical logics, multimodal logics, game theoretic semantics and linear logics, have the potential to cast new light on basic issues in the discussion of the unity of science.

This series provides a venue where philosophers and logicians can apply specific technical insights to fundamental philosophical problems. While the series is open to a wide variety of perspectives, including the study and analysis of argumentation and the critical discussion of the relationship between logic and the philosophy of science, the aim is to provide an integrated picture of the scientific enterprise in all its diversity.

For inquiries and submissions of proposals, authors can contact Christi Lue at christi.lue@springer.com

More information about this series at http://www.springer.com/series/6936

Christina Weiss
Editor

Constructive Semantics

Meaning in Between Phenomenology and Constructivism

Springer

Editor
Christina Weiss
Technical University of Darmstadt
Darmstadt, Germany

ISSN 2214-9775 ISSN 2214-9783 (electronic)
Logic, Epistemology, and the Unity of Science
ISBN 978-3-030-21312-1 ISBN 978-3-030-21313-8 (eBook)
https://doi.org/10.1007/978-3-030-21313-8

This Springer imprint is published by the registered company Springer Nature Switzerland AG
The registered company address is: Gewerbestrasse 11, 6330 Cham, Switzerland

Preface

The majority of this volume's articles grew out of the workshop *Constructive Semantics. Meaning in between Phenomenology and Constructivism* held at the University of Friedrichshafen between 30th of September and 1st of October 2016.[1] Its overall goal consisted in reflecting upon the possibility or even necessity to relate central concepts of constructive logic to phenomenology, that is to phenomenological concepts of meaning-constitution. Although the title might suggest discussions in between Husserlian phenomenology and Brouwerian intuitionism only, the range of considerations relating pragmatic and semantic aspects of knowledge was not at all restricted to Brouwer's concept of subjective construction on the one side, Husserl's concept of phenomenology as a study of transcendental consciousness on the other side. In contrast to such a limitation, the workshop explicitly sought for different approaches to phenomenology and construction, broadening the discussion towards more general epistemological questions of the form and function of phenomenology for pragmatism and pragmatism for phenomenology, especially in the light of recent developments in semantics, for instance, Robert Brandom's inferentialist account of meaning. Therefore, dialectical (Hegelian) concepts of phenomenology and pragmatics were likewise appreciated and were also put forward during the workshop.

The realisation of the workshop at the University of Friedrichshafen was financed by a sponsorship named Research Seed Capital (RiSC) of the Ministry of Science, Research and Arts of the state of Baden-Württemberg. Without this financial support, the workshop could not have been carried out.

Shahid Rahman, Editor of LEUS Springer, suggested the publication of this volume in the LEUS series. Many thanks to him for offering me this opportunity.

Cologne, Germany Christina Weiss

[1] See www.zu.de/veranstaltungen/2016/symposium-constructive-semantics.php.

Contents

Chapter 1
Introduction

Christina Weiss

Examining the relationship between the phenomenology of meaning-constitution on the one hand, constructivist accounts of (mathematical and logical) concepts on the other hand, is by no means new. During the years similarities between the two theoretical outlines have been explicated. Probably the most well known example in this direction consists in Oskar Becker's identification of Edmund Husserl's horizon-intentionality with L.E.J. Brouwer's concept of choice sequences (see Becker 1927). In recent times especially Mark van Atten has not only carved out similarities of Brouwerian and Husserlian concepts—especially with reference to the central role time consciousness plays in both approaches—but has actually achieved to transform obscure parts of Brouwer's general philosophical convictions into a transcendental phenomenological framework (see van Atten 2007, 2015).

New questions concerning the relationship of dialogical logic and dialogical constructivism in general to phenomenology are just gaining the attention they require, not least for a profound elaboration of dialogical constructivism itself.

The general interest, besides specific questions in between phenomenology and constructivism, having motivated the organization of the workshop *Phenomenology and Constructivism* and this collocation of articles, is to reflect on the concepts of phenomenology and constructivity themselves together with their respective interdependences.

Hereby we tie in with an idea, which Lothar Eley and others elaborate on in different occasions, aiming at what according to Eley might be called *Constructive Phenomenology* (see Eley 1985, 1976; Weiss 2018). Its specific turn consists in asking, in which respect a constructivist account of logic necessarily has to be accompanied by some sort of phenomenology for even gaining a concept of construction itself (see Eley 1985, p. XVI). In this respect, *Constructive Phenomenology's* interest doesn't solely consist in translating constructive terminology into phenomenological

C. Weiss (✉)
Institute of Philosophy, Technical University of Darmstadt, 64289 Darmstadt, Germany
e-mail: weiss@phil.tu-darmstadt.de

C. Weiss (ed.), *Constructive Semantics*, Logic, Epistemology, and the Unity of Science 44, https://doi.org/10.1007/978-3-030-21313-8_1

terminology, vice versa, but in developing a relational theory of what might be called *construction-with-appearance*.

The basic idea in this context might be summarized in the following way: The epistemological idea sitting in the back of the constructivist demand for a valid proof to be a proof of existence for an assumed proof-object is the phenomenological insight into the fundamental structure of *meaning-intention* and *meaning-fulfilment* (see Eley 1985, p. 17ff.). In other words: The phenomenological insight into the two-sided structure of *meaning-intention* and *meaning-fulfilment* illuminates constructivism's inherent epistemology. Constructivism necessitates this phenomenological element for its epistemological justification.

So this approach first of all follows the well known phenomenological interpretation of the distinction between *proposition and true proposition* as *meaning-intention* and *meaning-fulfilment* that Arend Heyting took from Becker's phenomenological interpretation of the idea of choice sequences for an intuitionistic account of logical junctors (see Eley 1985, 19ff.).

Now the crux of this account consists in conceptualizing the *fulfilment-aspect*, linguistically indicated by a deictic expression, as the presentational side of a concept. Fulfilling a conceptual content, in this view, becomes a necessary part of *concepts-as-meaning-intentions*. On the other hand, the fulfilling side, the spatio-temporal (self-)presentation of a concept, is constitutively bound to a linguistically mediated meaning-intention: *Every presentation is a presentation of something*. As a consequence, according to Eley, Husserl's (and Brouwer's) prioritization of intuition over signification has to be given up.[1] Instead of that, phenomenology should be conceptualized as consisting in the presentational, demonstrating role for a linguistically mediated intention. Simultaneously linguistically mediated meaning-intentions should be taken to function as guidelines for the demonstration-task.

This idea of phenomenology exposes a more semiotic character, in this respect it is closer to Peirce's considerations on the sign-character of thought. Therefore, it appears to be more apt to capture the form and function of phenomenology in and for constructive endeavours like dialogical logic or dialogical constructivism in general. On the other hand *Constructive Phenomenology* embodies what might be called a dialectical concept of phenomenology conceived as the presentational aspect of cognitive activity. This revitalization of a more Hegelian concept of phenomenology renders it possible to include genetical aspects of cognitive activity into its theoretical representation.[2]

[1] Eley discusses this partial turning back to a more Hegelian concept of phenomenology for instance with respect to Husserl's distinction of pre-predicative intentions and predicative forms. See (Eley 1969).

[2] In this volume especially Vojtěch Kolman and Christina Weiss pursue the road of a Hegelian understanding of phenomenology for clarifying the relationship between phenomenology and constructivism. Vojtěch Kolman pursues it in the shape of a dialectical account of *in itself-* and *for itself-* aspects of mathematical truth, explicated in a systematic-historic perspective of the self-unfolding of knowledge (see the further explications in this introduction and Kolman's article in this volume). Christina Weiss outlines some implications of a Hegelian reading of Kant's concept of schematism, emphasizing the profound change Hegel's concept of the self-appearance of knowledge has introduced into theorizing about the relationship between conceptual and intuitive form.

1.1 The Different Chapters

This volume's articles differ considerably with respect to their positions towards the necessity, possibility or even impossibility of integrating phenomenology and constructivism, as well as to the respective concepts of phenomenology and constructivity.

According to their goals and theoretical means they can be classified into three groups:

The *first group* includes the articles of Shahid Rahman, Mohammad Shafiei and Clément Lion. They all share a positive position towards an integration of phenomenology and constructivism. Especially Shafiei's and Lion's concept of phenomenology is of a Husserlian, transcendental-phenomenological shape. Additionally, their concept of constructivity in general, constructive logic in particular is the dialogical one that was brought forward by Paul Lorenzen and Kuno Lorenz, and has been further developed by Shahid Rahman and others. We have chosen to call this group of articles: *Integrating Transcendental Phenomenology into the Dialogical Framework*.

The *second group* of articles consists of two critical positions towards integrating phenomenology and contructivism, explicated by Mirja Hartimo and Claire Hill. Both share the reference point of Husserlian phenomenology as their basic critical means. They differ in their criticism's detailed content and direction of impact. Whereas Hartimo concentrates on a claimed incongruity between Husserl's notion of the phenomenological attitude and the concept of construction, of which she assumes to exist in the natural, unreduced attitude only, Hill directs attention to Husserl's classical conception of logic, which she is convinced of as being far more important in Husserl's work than his shift towards transcendental subjectivity. At least, Hill claims, is transcendental subjectivity bound to the classical non-subjective conception of logic, which underlines the latter's priority. We name this group of articles: *Critical Positions towards integrating Transcendental Phenomenology and Constructivism*.

The *third group* contains the articles of Vojtěch Kolman and Christina Weiss. Both conceptually share a Hegelian, dialectical conception of the relationship between constructivity and phenomenology. Kolman develops a dialectical account of the *Eleatic either-or-distinction*, which he transfers into a discussion of the relationship of truth and proof in mathematics, especially taking into account Gödel's incompleteness results. Weiss starts with a critical Hegelian reading of Kant's concept of *schematization* including his restriction of constructivity to mathematical constructivity. She then moves on to outline central aspects of what she calls *dialectical schematization*. Her argumentation closes with a criticism of Kuno Lorenz's dialogical constructivism, which in Weiss' conviction doesn't pay enough attention to the dialectical form of dialogical meaning-constitution. This group of articles is named: *Phenomenology and Constructivism as a Dialectical Relation*.

1.2 Specific Contents

In his article *Dialogues, Reasons and Endorsement* Shahid Rahman seeks to develop a dialogical framework of knowledge that has significant similarities to Robert Brandom's inferential pragmatism. What Rahman actually does is translating Brandom's idea of social rationality as *games of giving and asking for reasons* into the object-language of dialogical games. One could say that Rahman offers a differentiated pragmatic ground that is missing in Brandom's inferential semantics. At least he shows deep theoretical connections between the constructivist tradition and inferential semantics, which Brandom hasn't recognized until recently.[3]

Rahman firstly illustrates in which way dialogical games can be understood as *making explicit* games of giving and asking for reasons. He does so by interpreting Brandom's three conditions for the attribution of knowledge, which according to Rahman are:

1. Attribution of those commitments engaged by an assertion.
2. Attribution of those entitlements engaged by that assertion.
3. Endorsing the assertion and the commitments and entitlements attached to it (see Rahman in this volume), in the framework of dialogical logics as:

a. An obligation to the defensive move, that accompanies a player's bringing forward an assertion.
b. The right of the adversary to attack that assertion.
c. The so-called formal or Socratic rule (the Proponent can play an elementary proposition only if the Opponent has played it previously).

In the next step Rahman emphasizes that whereas dialogical logic already offers the means for making formally explicit "commitment", "entitlement" and "endorsement" it still lacks of explicating the reasons for an assertion in question. For this purpose, according to Rahman, it becomes necessary to integrate Martin-Löf's explicitation programme into dialogical logic more accurately. To illuminate the implications of integrating this explicitation programme into the very basics of dialogical logic is what Rahman seeks to pursue more precisely in his article. Again, the main title for Rahman's project is "Making it explicit". In this connection he focuses on the implementation of what he calls "local reasons" for a statement into the object language of dialogical games. The resulting dialogues of implementing local reasons into a dialogical game Rahman calls "dialogues for immanent reasoning". As a central characteristic of those local reasons for a statement Rahman claims their ability to embed content into a formal dialogical game.

"In fact, in principle; a local reason prefigures a material dialogue displaying the content of the proposition stated. This aspect makes up the ground level of

[3]See concerning Brandom's surprising ignorance of similarities between his inferential conception of meaning and the meaning-conception of methodical constructivism Kambartel (2008) and Brandom (2008). The same amazing *non-consideration* also holds for Brandom's treatment of phenomenology. *Tales of the mighty dead* (Brandom 2002) manages to deal with historic-systematic problems of intentionality without even mentioning Edmund Husserl.

the normative approach to meaning of the dialogical framework, in which *use*—or dialogical interaction—is to be understood as *use prescribed by a rule* [...]" (Rahman in this volume, p. 42).

So, a local reason is meant to function as an indicator for a material, that is, content-dependent dialogue, a content-dependent dialogical interaction. Inferentialist Robert Brandom defines the so-called *material inference* as an inference that is true, not because of a formal inference-rule it obeys, but because of the material truth, that is truth with regards to what Brandom calls *conceptual content. Conceptual content* is meant to signify relations of conceptual inclusion and exclusion instantiated in and by everyday use:

Identifying something as a dog includes its being a mammal and excludes its being a cat.

The problem with this—concerning its form—Carnapean definition of materiality consists in the fact that the information-theoretical idea to identify the logical inclusiveness/exclusiveness of a statement with its content, originally proposed by Bar-Hillel/Carnap, is bound to a formalized logical language of first order relations (see Bar-Hillel/Carnap 1964). Instead of introducing materiality, for instance by means of a phenomenological analysis, Brandom introduces it with the help of formal semantics. His concepts of *conceptual content* and *material inference* thus remain formal concepts, even if Brandom claims the opposite.

An interesting question one might ask with respect to Rahman's concept of *local reason* would be, how the concept of local reason could be implemented into a general semantic framework, for example into dialogical constructivism of Kuno Lorenz's shape: Which exact role does *local reason* play beyond, or better prior to the explicit formulation of the semantics of logical connectives? Does *local reason,* for instance, function as a bridge between what Lorenz calls *singularizing instantiation* of a schema and *universalizing adduction* of a single action? At least, many further considerations are obviously possible based on Rahman's discussion of *local reasons.*

Mohammad Shafiei's paper *A phenomenological Analysis of the Distinction between Structural Rules and Particle Rules in Dialogical Logic* aims at relating architectonical invariants of dialogical logic like the distinction of structural rules and particle rules to Husserl's distinction between consequence-logic and truth-logic.

For this purpose Shafiei firstly introduces the well-known Husserlian distinction between consequence-logic and truth-logic emphasizing Husserl's position that problems related to logical consequence and consistency (consequence-logic) are problems of logical form only, in contrast to questions concerning, for instance, the relationship between the truth of logical antecedents and the truth of logical consequents. Shafiei continuously stresses that consequence-logic and truth-logic should not be regarded as two sorts of logic, but as two viewpoints on two aspects of logic. For deepening the understanding of Husserl's distinction between consequence-logic and truth-logic Shafiei relates these two forms of logical reasoning to the phenomenological idea of acts of *meaning-intention* and acts of *meaning-fulfilment.* Whereas the consequence-level is concerned with relations, formal inclusion or exclusion, among meaning-intentions, the truth-level is concerned with aspects of the fulfilment of meaning-intentions. Shafiei adds that truth, that is, the fulfilment-level, is always an experience of truth and therefore has to take into account the peculiarities of experience. Therefore, Shafiei argues that it is necessary to distinguish and not confuse the different sorts of acts involved on the consequence-level and on the truth-level.

Shafiei aims at offering a more technical framework for Husserl's distinction between consequence-level and truth-level and finds it in the works of dialogical logic, developed by Lorenzen, Lorenz, Rahman and others. He firstly introduces the distinction between structural rules, governing the dialogue as a whole and particle rules, governing specific moves. Secondly he relates the distinction of structural rules and particle rules to the distinction of *play level* (permitted moves) and *strategy level* (a chain of permitted moves in order to establish a proof, if possible).

Finally Shafiei arrives at the conclusion that Husserl's distinction between consequence-level and truth-level is mirrored in the distinction of particle rules, which are concerned with the introduction of the different logical connectives, and structural rules, which imply rules for the dialogue as a whole. Furthermore, Shafiei criticizes truth-functional and model-theoretic approaches for confusing meaning-intentions and meaning-fulfilment.

Clément Lion's paper *A dialogical account of the intersubjectivity of intuitionism* offers an outline of integrating a phenomenologically revised version of Brouwer's intuitionism on the one hand, dialogical semantics brought forward by Paul Lorenzen, Kuno Lorenz, Shahid Rahman and others on the other hand. His main concern in this context is an intersubjective re-interpretation of Brouwer's concept of the *Creative Subject.*

Lion opens the discussion by asking the question, if and in which respect occasional, subject-dependent expressions could be regarded as belonging to the mathematical universe of discourse. He then decides to choose Brouwer's famous concept of *free choice sequences* as a persuasive candidate for such an appearance of subject-dependent expressions in mathematical reasoning.

A crucial concept that Lion discusses in relation to mathematical objects is their supposed ideality. He tries to work out intersubjective foundations for the concept of ideal meaning. Lion's model of the problem of an intersubjective foundation of the ideality of meaning is a communicative model consisting of a *speaker role* and a *hearer role*.

Now it is clear that if ideality is conceptualized through the eyes of communication theory it becomes a problem of transmission, that is transmission of an original (ideal) meaning the speaker intends, to the hearer, who constitutively cannot share the speaker's original intention. Precisely because of the hearer's lack of original meaning-intention a play of asking the speaker for necessary specifications comes into being.

In this connection Lion critically discusses Husserl's shift towards an eidetic analysis of *horizon-intentionality*, which according to Lion doesn't offer any significant role for contingent factuality and therefore isn't apt to account for occasional expressions. Therefore, Lion turns towards O. Becker's concept of transcendental reduction without eidetic reduction. He mentions Becker's Heideggerian turn, Becker's conception of *free choice sequences* "as being anchored in the factual individual existences, which project themselves forward into the full obscurity of the future" (Lion, p. 105 f. in this volume).

Lion proceeds in discussing the problem of intersubjectivity. He re-interprets a sentence of Oskar Becker, stating the necessity of the *factual other* because of an assumed weakness of the power of imagination.

"It follows that there is a place for deictic expressions at the very basis of mathematics itself, if mathematically acting subjects in front of me are carrying out operations which I would absolutely not have thought of by myself and which constitute the fundamental basis of my own mathematical activity and development" (Lion, p. 109 in this volume).

Lion tends to equalize the meaning of an expression with a subjective idea of it. He links the assumed existence of occasional expressions inside mathematics with the necessary existence of intersubjectivity for mathematical thinking and learning.

"That means that the incomplete perceptible object that the mathematician points to, for example on a blackboard, saying "behold what I have constructed", would not be communicable through the similarity of "experienced world" for both the speaker and the hearer, but rather through the act of providing, through speech acts, the precisions which are requested by the hearer to whom it has been addressed" (Lion, p. 110 f. in this volume).

An interesting further question in this context would be, how to understand the mathematician's "pointing to an object"? In the line of Husserlian Phenomenology one would have to take into account the distinction of *sensible intuition* and *categorial intuition* for a precise analysis of the meaning of "pointing to an object".

Lion finally interprets the supposed ideal meaning of mathematical objects in a dialogical way by taking the interactive specifications of a possibly infinite horizon of determinations to represent the ideal unity itself.

Mirja Hartimo's article *Constitution and Construction* advances a critical position on the proposal of interpreting transcendental phenomenology in a constructivist way. With respect to central notions in question, she argues for an incompatibility of Husserl's notion of constitution on the one hand, the mathematical activity of construction on the other hand. Hartimo develops her argument around the Husserlian distinction of *natural attitude* and *transcendental-phenomenological* attitude. Her criticism against efforts of re-interpreting Husserl's transcendental-phenomenological conception of constitution through constructivist eyes, which for example, Mark van Atten brings forward by reformulating Brouwer's concept of the *Creative Subject* in terms of Husserl's transcendental subject, basically runs along the lines of carving out a fundamental misunderstanding with respect to the role of *constitution*, which is carried out by transcendental consciousness. Her main point consists in emphasizing that the terms *constitution* and *construction*, according to Husserl, belong to different attitudes. Whereas *construction* in Husserl's view is rooted in the *natural attitude*, *constitution* is a term meaningful only in the *phenomenological attitude*. Identifying *constitution* and *construction* in Hartimo's eyes leads to a confusion of the role of the different attitudes in Husserlian phenomenology and accordingly to a misunderstanding of the phenomenological enterprise.

Hartimo starts her discussion by referring to two different understandings of the notion of construction, which Husserl himself explicated. The first one contains a rather broad understanding of construction in the sense of building any mathematical theory whatsoever. According to this version of construction, theoretical mathematical activity is constructive, independent of specific attitudes towards foundational questions, like the *tertium non datur* or the introduction of the real numbers.

Husserl's second understanding of construction according to Hartimo in a narrower sense refers to "an activity carried out in the 'theory of judgment'" (Hartimo, p. 123 in this volume).

To elucidate her critical position Hartimo distinguishes sharply between the different functional roles of the two attitudes in and for phenomenology. Whereas existing things, their different determinations and relationships belong to the realm of the *natural attitude*, their meanings, their forms of givenness belong to the realm of the *phenomenological attitude*, of the phenomenological uncovering of the invariant eidetic forms of meaning-constitution.

Now, one could object, as Hartimo points out, that because in transcendental phenomenology mathematics and phenomenology are considered as eidetic and not as empirical sciences, the objects of mathematical thought as well as the objects of the phenomenologically reduced transcendental consciousness don't belong to the realm of *natural attitude*. One might conclude that, because mathematics and transcendental phenomenology both are about essences and not about facts, their subject should be the same or at least similar. But this conclusion in Hartimo's eyes is profoundly mistaken because of the following reasons:

Mathematics and phenomenology deal with completely different forms of essences, according to Husserl. Whereas mathematical essences are supposed to be exact and ideal, phenomenological essences are regarded as unexact and morphological (see Hartimo, p. 126 in this volume).

And whereas mathematical essences are the essences of transcendent objects, phenomenological essences are essences immanent in consciousness.

"Phenomenology and mathematics thus belong to crucially different dimensions that should not be conflated" (Hartimo, p. 127 in this volume).

Hartimo strictly sticks to the Husserlian distinctions of *exactness/morphological* and *immanence of consciousness/transcendence of objects*. An interesting question in this connection is: Does constructivist mathematics through developing a completely different view concerning the meaning of mathematical entities challenge Husserl's basic distinctions with respect to the relationship of mathematics and phenomenology itself? It might well be that constructivists like Brouwer did have in mind a sort of phenomenological mathematics that could prove to overcome classical separations of *exact/not exact* or *immanence/transcendence*. If this was the case, there would indeed be a chance of integrating constructivism and phenomenology, whereby phenomenology as well as constructive mathematics would undergo some significant changes, of course.

Claire Hill's article on *Husserl's Purely Logical Chastity Belt* brings forward a likewise critical position with respect to the possibility of integrating constructivism and Husserlian phenomenology. Though Hill states at the beginning that her considerations aren't about constructivism nor anti-constructivism, she indirectly unfolds a critical position towards identifying constructivist ideas with Husserlian philosophy. In this connection it is her central concern to emphasize that the elaboration of transcendental phenomenology is only one part of Husserl's oeuvre. The other part, according to Hill, consists of Husserl's research into the foundations of pure logic itself, that is, foundational research without the simultaneous transcendental-phenomenological perspective. The argument Hill develops between the lines, based on the supposed distinction between Husserl's logical and Husserl's transcendental-phenomenological investigations, against identifying Husserl as a constructivist might be summarized as follows: Because Husserl does advocate for the ideality of logic, its ideal order, existing independent of concrete subjects and their concrete acts of intentionality, transcendental phenomenology in its shape as well as in its contents mustn't be regarded as a science of autonomous subjectivity, but as constitutively bound to a presupposed logical ideality. Hill doesn't say this explicitly, but this kind of thought seems to sit in the back of her considerations. "[…] That subjective, transcendental logic had to find its complement in pure, objective, a priori formal logic, free from acts, subjects, or empirical persons or objects belonging to actual reality and entirely grounded in conceptual essentialities, and vice versa" (Hill, p. 137 in this volume).

The enterprise of phenomenology taken as a study of the performances of consciousness coming along with ideal meanings, in this respect might be interpreted as a bridge, a third, relating ideal meanings and subjective constitutions in a rationally convincing manner.

Now with respect to constructivism, the issue of the applicability of constructivism to phenomenology, phenomenology to constructivism, one might object that Brouwer's ideal *Creative*, constructive *Subject* doesn't necessarily contradict Husserl's complementary view of subjective constitution and ideal meaning. Mark van Atten's work, for example, in reformulating the intuitionistic foundations through transcendental-phenomenological means, doesn't aim at neglecting the ideality of logical meanings, but at explicating constructive analogies of mathematical and logical ideality (see van Atten 2007, 2015). But Hill's general point of course remains powerful and opens up further questions concerning the interdependency of concepts of subjectivity and concepts of meaning.

In *The Truth of Proof: A Hegelian Perspective on Constructivism* Vojtěch Kolman adopts a different position on the very concept of constructivity and phenomenology likewise. Instead of relating phenomenological considerations of the Husserlian shape to constructivist considerations of Brouwer or Heyting Kolman takes a Hegelian perspective on questions regarding the relationship of *truth* and *proof.* In this respect he combines different ideas. Firstly, he takes Hegel's insight into the progressive 'nature' of knowledge and truth and the interrelated relationship of *being in itself* and *being for us* to gain a deeper understanding of the concept of constructivity itself. In this connection he challenges ahistorical, purely structural conceptions of constructivity. Secondly, Kolman applies his reformed concept of constructivity to the split of *knowledge* and *truth* (provoked by Gödel's theorems), which he seeks to understand as two co-dependent dialogical and dialectical roles.

Kolman opens up his considerations with a historical discussion of the differing understanding of the exclusive *either-or-distinction* in canonical philosophical endeavours, starting with the Eleatics followed by Plato and Kant, arriving at Hegel.

He introduces the four philosophical traditions he refers to in the form of two tandems: The Eleatics and Plato in the first place, secondly Kant and Hegel.

Concerning the first tandem, the Eleatics and Plato, Kolman emphasizes that whereas the Eleatics introduced the general idea of proof by contradiction and the simultaneous ontologization of the either-or-dichotomy, Plato challenged the Eleatic method of indirect proof by showing that a contradiction found in a statement *A* doesn't necessarily lead to the truth of *non-A*, because this might be shown to contain a contradiction, too. Kolman stresses that the Platonic reading of the purely formal understanding of either-or by the Eleatics implies a demand to a clarification of the concepts in question. That is, the heuristic role of showing the possibility of falsity of both, *A* and *non-A*, in Plato's dialogues consists in the demonstrative function with regard to a lack of clarity of the respective concepts.

The second tandem begins with Kant's accentuation of the relativity of all truth and meaning with respect to our cognitive faculties. Kolman views this relativity, this *true for us*, as the reason, why Kant demanded that mathematical objects should be constructions in intuition. He stresses that in spite of Kant's constructive enlargement of cognition through his claim of the necessary givenness of any object in intuition, Kant doesn't clarify the exclusive either-or-distinction but even multiplies it by separating the conceptual and the intuitive realm as well as the appearance an object has *for us* from its supposed *being in itself*. Kolman regards this formalistic copying of either-or, regardless of the content of application, as the key source and goal of Hegel's criticism of Kant's philosophy.

He then moves on to apply Hegel's dialectical version of *in itself/for us*, in which the different aspects in contrast to Kant's view are regarded as relational with respect to the different stations of the self-revealing process of knowledge, to questions of number theory.

"The given lesson is that yesterday's 'in itself' is today's 'for itself' and that this is not a deficit but an inherent pattern of knowledge which, so to speak, must construct itself from its own resources" (Kolman, p. 154 in this volume).

Comparing this dialectical procedural relationship of *in itself/for itself* with constructivist thinking, Kolman carves out that the Hegelian concept of dialectical knowledge shares with constructivist concepts a critical view of a purely formal, that is Eleatic understanding of either-or: "Put differently, the constructivist reading of Either-Or does not allow one to proceed directly from the exclusion of the Either to the confirmation of the Or because there is typically a lot of work still to be done" (Kolman, p. 154 in this volume).

But on the other hand, and in Kolman's argumentation more important, he stresses the one-sidedness of especially Brouwer's considerations with respect to the *continuum* and *free choice sequences*. In Kolman's eyes Brouwerian mathematics fails to account for the *in itself* of mathematical objects, for the necessary complement of the subjective *for us*. That failure finally opens the door for solipsistic arbitrariness.[4]

From the problematic results of Brouwer's subjectivism Kolman proceeds to discuss different approaches with respect to the relation of subjective and objective aspects in mathematical reasoning. One of the central outcomes of this discussion is the role he attributes to Gödel's incompleteness results. He claims that Gödel by his famous proof of non-provability of a true mathematical sentence for every rich enough formal axiomatization has dialectically revisioned the *either-or-distinction* of *truth* and *proof* by proving that something, a particular sentence, is improvable, but true. "The truth of the mechanical concept of proof turned out to be another, broader concept of proof which is to be, consequently with the very notion of truth. The concepts of proof and truth are the new horns of the old Either-Or-dichotomy, this time arising within the subject itself by means of its ability to reflect on itself, i.e. to be self-conscious" (Kolman, p. 161 in this volume).

After proceeding the discussion on Gödel's results with a discussion of the *ω-rule*, Lorenzen's distinction of *full-* and *semi-formalism* with respect to the *ω-rule* and Cantor's diagonalization result of the non-denumerability of the reals, Kolman finally arrives at Lorenzen's dialogical semantics and Brandom's rational pragmatics. According to Kolman their key achievement is the uncovering of the social nature of reasoning.

In *Constructive Semantics: On the necessity of an appropriate concept of schematization* Christina Weiss discusses the general form and relevance of *constructivity* in and for philosophical theory. Her overall goal in this respect is a demonstration of the necessity of what she calls a *dialectical concept of schematizing* semantic and pragmatic aspects of knowledge.

The article is subdivided into three parts. In the first part Weiss critically reconstructs Immanuel Kant's ideas on *transcendental schematism.* Her main points of criticism against Kant's conception of the schematizing function, that in the *Critique of Pure Reason* is meant to integrate the conceptual and the intuitive realm, is firstly the abstracting method by which Kant arrives at *pure time* as being *homogeneous* with both, pure concept and empirical appearance, therefore representing the mediating function between pure concept and empirical intuition. She argues that in conceiving the trias of pure concept, pure time and empirical appearance along the lines of a traditional genus-species-tree Kant neither clarifies the important concept of *homogeneity* he uses for a derivation of pure time as *the* mediating-function between pure concept and empirical appearance nor does he capture the actual relationship between the aspects in question. This renders Kant's account of schematism arbitrary and unsatisfactory. This is of special importance, because Kant's method makes it impossible to understand the process of schematization other than as a psychological enigma.

[4]This is of course a strong claim, especially if one takes into consideration Mark van Atten's work, explicating homogeneities between Brouwerian constructivism and Husserlian phenomenology. At least it should renew the discussion of a possibly similar arbitrariness in Husserlian phenomenology.

Weiss emphasizes that in Kant's formulation there already exists a different, dialectical account of the relationship of pure concept, pure time and empirical intuition: "Now one could say that time is related to categories as well as to appearances in the sense that time, as the a priori intuition of the manifold is the 'whereof' of the synthetic unity, which the categories constitute on the one side, the 'Being-in' of every concrete appearance on the other side (Weiss, p. 176 in this volume).

With the result of Kant's insufficient concept of schematization Weiss moves on to formulate the problem of schematization from a Hegelian perspective. In this context she focuses on the fundamentally different conception of unity Hegel exposes. Whereas Kant's concept of the unity of apperception remains a psychologistic connection of two different faculties, reason and intuition, Hegel's concept of unity is a relational unity of conceptual and presentational aspects of knowledge. Through reconstructing schematization as a dialectical process of *form and content*, *concept and intuition* Hegel arrives at profoundly different results concerning the different notions in question like *concept* and *intuition*, but also concerning knowledge as a unity. *Unity* in Hegel's eyes is a dynamic unity of semantic and presentational aspects of knowledge, that is, a symbolic unity.

In a last step Weiss uses the results of her Hegelian reading of Kant to expatiate on some assumed deficiencies of Kuno Lorenz's concept of dialogical construction. Her main criticisms consists in what she calls an *underestimation of the dialectics* sitting in Lorenz's concept of the *elementary learning and teaching situation*.

Her conclusion is that a more precise account of the dialectical relationship between *singularizing instantiation* of a schema and *universalizing adduction* of singular action leads to the necessity of strengthening and re-formulating the role of phenomenology in and for dialogical constructivism.

References

Bar-Hillel, Y., Carnap, R. (1964). An outline of a theory of semantic information. In: Bar-Hillel, *Language and Information*. Reading (pp. 221–274). Addison-Wesley.

Becker, O. (1927). Mathematische Existenz. Untersuchungen zur Logik und Ontologie mathematischer Phänomene. *Jahrbuch für Philosophie und phänomenologische Forschung, 8*, 439–809.

Brandom, R. (2002). *Tales of the mighty dead: Historical essays in the metaphysics of intentionality*. Cambridge: Harvard University Press.

Brandom, R. (2008). *Responses*. In: Stekeler-Weithofer, pp. 209–229.

Eley, L. (1969). *Metakritik der formalen Logik. Sinnliche Gewißheit als Horizont der Aussagenlogik und elementaren Prädikatenlogik*. Den Haag: Martinus Nijhoff.

Eley, L. (1976). *Hegels Wissenschaft der Logik*. München: Wilhelm Fink.

Eley, L. (1985). *Philosophie der Logik*. Darmstadt: Wissenschaftliche Buchgesellschaft.

Kambartel, F. (2008). Meaning, justification and truth. In: Stekeler-Weithofer, P., *The pragmatics of making it explicit* (pp. 99–108). Amsterdam: John Benjamins.

van Atten, M. (2007). *Brouwer meets Husserl. On the phenomenology of choice sequences*. Dordrecht: Springer.

van Atten, M. (2015). Construction and constitution in mathematics. In M. van Atten (Ed.), *Essays on Gödel's reception of Leibniz, Husserl and Brouwer* (pp. 237–288). Switzerland: Springer.

Weiss, C. (2018). Towards a phenomenology of schematization. *Cybernetics and Human Knowing, 24*, 245–260.

Part I
Integrating Transcendental Phenomenology into the Dialogical Framework

Chapter 2
Dialogues, Reasons and Endorsement

Shahid Rahman

Abstract The main aim of the present paper is to show that, if we follow the dialogical insight that reasoning and meaning are constituted during interaction, and we develop this insight in a dialogical framework for Martin-Löf's Constructive Type Theory, a conception of knowledge emerges that has important links with Robert Brandom's (1994, 2000) *inferential pragmatism*. However, there are also some significant differences that are at center of the dialogical approach to meaning. The present paper does not discuss explicitly phenomenology, however, one might see our proposal as setting the basis for a further study linking phenomenology and the dialogical conception of meaning—the development of such a link is part of several ongoing researches.

Keywords Dialogical logic · Transcendental logic · Truth · Consequence · Structural rules · Particle rules · Phenomenology · Intentionality · Play-level · Strategy-level

2.1 Introduction

The present paper aims at showing that, if we follow the dialogical insight that reasoning and meaning are constituted during interaction, and we develop this insight in a dialogical framework for Martin-Löf's Constructive Type Theory (CTT)°, a conception of knowledge emerges that has important links with Robert Brandom's (1994, 2000) *inferential pragmatism*.

S. Rahman (✉)
Université de Lille, Philosophie, CNRS-UMR: 8176, Pont de Bois BP, 59653 Villeneuve d'Ascq, France
e-mail: shahid.rahman@univ-lille.fr

C. Weiss (ed.), *Constructive Semantics*, Logic, Epistemology, and the Unity of Science 44, https://doi.org/10.1007/978-3-030-21313-8_2

Indeed according to Brandom (see for example, 2000, Chap. 3) attribution of knowledge as determined by *games of giving and asking for reasons* is dependent upon three main conditions[1]

1. Attribution of those commitments engaged by an assertion
2. Attribution of those entitlements engaged by that assertion
3. Endorsing the assertion and the commitments and entitlements attached to it.

Our task now lies in developing *games of giving and asking for reasons* where some specific moves make explicit the fulfilment of the conditions mentioned above. In fact the dialogical framework already can be seen as displaying such kind of moves in the following way

1. Commitment corresponds to the defensive move that one player is obliged to, when bringing forward some assertion
2. Entitlement correspond to the right the adversary to attack that assertion
3. Endorsement corresponds to the so-called formal rule (also known as the Socratic rule).

Actually, as discussed further on in the present paper, in some recent talks Martin-Löf offered some insightful reflections on the contribution of the dialogical approach to the deontic and epistemic interface. More precisely, in his Oslo and Stockholm lectures, Martin-Löf's (2017a, b) condenses the dialogical view on commitments and entitlements that he declines respectively as on one hand *must-requests* (commitments or obligations) and on the other *may-requests* (or entitlements or rights) as follows[2]:

> [...] So, let's call them rules of interaction, in addition to inference rules in the usual sense, which of course remain in place as we are used to them.
>
> [...] Now let's turn to the request mood. And then it's simplest to begin directly with the rules, because the explanation is visible directly from the rules. So, the rules that involve

---

[1] The relation between dialogical logic and the games of asking and giving reasons has already been pointed out by (Keiff 2007) and (Marion 2006, 2009, 2010). See for example:

> My suggestion is simply that dialogical logic is perfectly suited for a precisification of these 'assertion games'. This opens the way to a 'game-semantical' treatment of the 'game of giving and asking for reasons': 'asking for reasons' corresponds to 'attacks' in dialogical logic, while 'giving reasons' corresponds to 'defences'. In the Erlangen School, attacks were indeed described as 'rights' and defences as 'duties', so we have the following equivalences:
>
> Right to attack $\leftrightarrow$ asking for reasons
>
> Duty to defend $\leftrightarrow$ giving reasons
>
> The point of winning 'assertion games', i.e., successfully defending one's assertion against an opponent, is that one has thus provided a justification or reason for one's assertion. Referring to the title of the book [Making it Explicit], one could say that playing games of 'giving and asking for reasons' implicitly presupposes abilities that are made explicit through the introduction of logical vocabulary (Marion 2010, p. 490).

[2] Ansten Klev's transcription of Martin Löf (2017a, pp. 1–3, 7).

request are these, that if someone has made an assertion, then you may question his assertion, the opponent may question his assertion.

$$(Req1)\frac{\vdash C}{?\vdash_{may} C}$$

Now we have an example of a rule where we have a may. The other rule says that if we have the assertion $\vdash C$, and it has been challenged, then the assertor must execute his knowledge how to do $\vdash C$. […].

$$(Req2)\frac{\vdash C \quad ?\vdash C}{\vdash_{must} C'}$$

In relation to the third condition of Brandom, *endorsement*, it involves the use of assertions brought forward by the interlocutor. In this context Sundholm (2013, p. 17) produced the following proposal that embeds Austin's remark (Austin 1946, p. 171) on assertion acts in the context of inference:

When I say therefore, I give others my authority for asserting the conclusion, given theirs for asserting the premisses.

Herewith, the assertion of one of the interlocutors *entitles* the other one to endorse it. Moreover, in recent lectures, Martin-Löf (2015) used this dialogical perspective in order to escape a form of circle threatening the explanation of the notions of inference and demonstration. A demonstration may indeed be explained as a chain of (immediate) inferences starting from no premisses at all. That an inference

$$\frac{J_1 \ldots J_n}{J}$$

is valid means that one can make the conclusion (judgement J) evident on the assumption that $J_1 \ldots J_n$ are known. Thus the notion of epistemic assumption appears when explaining what a valid inference is. According to this explanation however, we cannot take 'known' in the sense of *demonstrated*, or else we would be explaining the notion of inference in terms of demonstration when demonstration has been explained in terms of inference. Hence the threatening circle. In this regard Martin-Löf suggests taking 'known' here in the sense of *asserted*, which yields epistemic assumptions as judgements others have made, judgements whose responsibility others have already assumed. An inference being valid would accordingly mean that, given others have assumed responsibility for the premisses, I can assume responsibility for the conclusion.

Thus, when explaining the notion of immediate inference we are assuming, not that the premisses have been demonstrated but that they have been asserted by someone else, and this can be endorsed. In a dialogical setting we are thus imagined as acquiring certain knowledge on trust of the Opponent's assertions, on which basis the Proponent may make evident certain further pieces of knowledge. In this solution to the circularity threatening the explanations of demonstration and immediate inference Martin-Löf understands epistemic assumption as assertoric assumption.

This goes in hand with Martin-Löf's further point that in the dialogical setting, the deontic force has priority over other layers.

Thus, the dialogical framework already seems to offer a formal system where the main features of Brandom's epistemological games can be rendered explicit. However, the system so far does not make explicit the reasons behind an assertion. In order to do so we need to incorporate into the dialogical framework expressions standing for those reasons. This requires combining dialogical logic with Per Martin Löfs Constructive Type Theory (1984) in a more thorough way.

We call the result of such enrichment of the expressive power of the dialogical framework, *dialogues for immanent reasoning* precisely because *reasons* backing a statement, now *explicit* denizens of the object-language of plays, are *internal* to the development of the dialogical interaction itself—see Rahman et al. (2018).[3]

However, despite the undeniable links of the dialogical framework to both CTT and Brandom's inferentialist approach to meaning there are also some significant differences that are at the center of the dialogical conception of meaning, namely the identification of a level of meaning, i.e. the play-level, that does not reduce to the proof-theoretical one. We will start by presenting the main features of dialogues for immanent reasoning and then we will come back to the general philosophical discussion on the play-level as the core of what is known as *dialogue-definiteness*.

The present paper does not discuss explicitly phenomenology, however, Shafiei (2017) developed in his thesis: *Intentionnalité et signification: Une approche dialogique*, a thorough study of the bearing of the dialogical framework for phenomenology. Nevertheless, his work did not deploy the new development we call *immanent reasoning*. So, one might see our proposal as setting the basis for a further study linking phenomenology and the dialogical conception of meaning.

2.2 Local Reasons

Recent developments in dialogical logic show that the Constructive Type Theory approach to meaning is very natural to the game-theoretical approaches in which (standard) metalogical features are explicitly displayed at the object language-level.[4] This vindicates, albeit in quite a different fashion, Hintikka's plea for the fruitfulness of game-theoretical semantics in the context of epistemic approaches to logic, semantics, and the foundations of mathematics.[5]

From the dialogical point of view, the actions—such as choices—that the particle rules associate with the use of logical constants are crucial elements of their full-

[3]In fact, the present paper relies heavily on the main technical and philosophical results of Rahman et al. Clerbout (2018). However, some important modifications have been introduced, particularly in the conception of strategic reasons. Many thanks to the reviewers of the present paper, I owe important modifications to their suggestions.

[4]Such as developed in Rahman et al. (2018) and also in Clerbout and Rahman (2015).

[5]Cf. Hintikka (1973).

fledged (local) meaning: if meaning is conceived as constituted during interaction, then all of the actions involved in the constitution of the meaning of an expression should be made explicit; that is, they should all be part of the object-language.

This perspective roots itself in Wittgenstein's remark according to which one cannot position oneself outside language in order to determine the meaning of something and how it is linked to syntax; in other words, language is unavoidable: this is his Unhintergehbarkeit der Sprache. According to this perspective of Wittgensteins, language-games are supposed to accomplish the task of studying language from a perspective that acknowledges its internalized feature. This is what underlies the approach to meaning and syntax of the dialogical framework in which all the speech-acts that are relevant for rendering the meaning and the "formation" of an expression are made explicit. In this respect, the metalogical perspective which is so crucial for model-theoretic conceptions of meaning does not provide a way out. It is in such a context that Lorenz writes:

> Also propositions of the metalanguage require the understanding of propositions, [...] and thus cannot in a sensible way have this same understanding as their proper object. The thesis that a property of a propositional sentence must always be internal, therefore amounts to articulating the insight that in propositions about a propositional sentence this same propositional sentence does not express a meaningful proposition anymore, since in this case it is not the propositional sentence that is asserted but something about it.
>
> Thus, if the original assertion (i.e., the proposition of the ground-level) should not be abrogated, then this same proposition should not be the object of a metaproposition [...].[6]
>
> While originally the semantics developed by the picture theory of language aimed at determining unambiguously the rules of "logical syntax" (i.e. the logical form of linguistic expressions) and thus to justify them [...]—now language use itself, without the mediation of theoretic constructions, merely via "language games", should be sufficient to introduce the talk about "meanings" in such a way that they supplement the syntactic rules for the use of ordinary language expressions (superficial grammar) with semantic rules that capture the understanding of these expressions (deep grammar).[7]

Similar criticism to the metalogical approach to meaning has been raised by Sund holm (1997, 2001) who points out that the standard model-theoretical semantic turns semantics into a meta-mathematical formal object in which syntax is linked to meaning by the assignation of truth values to uninterpreted strings of signs (formulae). Language does not express content anymore, but it is rather conceived as a system of signs that speak about the world—provided a suitable metalogical link between the signs and the world has been fixed.

Ranta (1988) was the first to link game-theoretical approaches with CTT. Ranta took Hintikka's (1973) Game-Theoretical Semantics (GTS) as a case study, though his point does not depend on that particular framework: in game-based approaches, a proposition is a set of winning strategies for the player stating the proposition. In game-based approaches, the notion of truth is at the level of such winning strategies. Ranta's idea should therefore in principle allow us to apply, safely and directly,

---

[6]Lorenz (1970, p. 75), translated from the German by Shahid Rahman.

[7]Lorenz (1970, p. 109), translated from the German by Shahid Rahman.

instances of game-based methods taken from CTT to the pragmatist approach of the dialogical framework.

From the perspective of a general game-theoretical approach to meaning however, reducing a proposition to a set of winning strategies is quite unsatisfactory. This is particularly clear in the dialogical approach in which different levels of meaning are carefully distinguished: there is indeed the level of strategies, but there is also the level of plays in the analysis of meaning which can be further analysed into local, global and material levels. The constitutive role of the play level for developing a meaning explanation has been stressed by Lorenz in his (2001) paper:

> Fully spelled out it means that for an entity to be a proposition there must exist a dialogue game associated with this entity, i.e., the proposition A, such that an individual play of the game where A occupies the initial position, i.e., a dialogue D(A) about A, reaches a final position with either win or loss after a finite number of moves according to definite rules: the dialogue game is defined as a finitary open two-person zero-sum game. Thus, propositions will in general be dialogue-definite, and only in special cases be either proof-definite or refutation-definite or even both which implies their being value-definite.
>
> Within this game-theoretic framework [...] truth of A is defined as existence of a winning strategy for A in a dialogue game about A; falsehood of A respectively as existence of a winning strategy against A.[8]

Given the distinction between the play level and the strategy level, and deploying within the dialogical framework the CTT-explicitation program, it seems natural to distinguish between local reasons and strategic reasons: only the latter correspond to the notion of proof-object in CTT and to the notion of *strategic-object* of Ranta. In order to develop such a project we enrich the language of the dialogical framework with statements of the form "$p : A$". In such expressions, what stands on the left-hand side of the colon (here p) is what we call a local reason; what stands on the right-hand side of the colon (here A) is a proposition (or set).

The local meaning of such statements results from the rules describing how to compose (synthesis) within a play the suitable local reasons for the proposition A and how to separate (analysis) a complex local reason into the elements required by the composition rules for A. The synthesis and analysis processes of A are built on the formation rules for A.

The most basic contribution of a local reason is its contribution to a dialogue involving an elementary proposition. Informally, we can say that if the Proponent **P** states the elementary proposition A, it is because **P** claims that he can bring forward a reason in defence of his statement, it is this reason that provides content to the proposition.

[8] Lorenz (2001, p. 258).

2.2.1 Local Meaning and Local Reasons

Dialogues are games of giving and asking for reasons; yet in the standard dialogical framework, the reasons for each statement are left implicit and do not appear in the notation of the statement: we have statements of the form **X** ! A for instance where A is an elementary proposition. The framework of dialogues for immanent reasoning allows to have explicitly the reason for making a statement, statements then have the form **X** a:A for instance where a is the (local) reason **X** has for stating the proposition A. But even in dialogues for immanent reasoning, all reasons are not always provided, and sometimes statements have only implicit reasons for bringing the proposition forward, taking then the same form as in the standard dialogical framework: **X** !A. Notice that when (local) reasons are not explicit, an exclamation mark is added before the proposition: the statement then has an implicit reason for being made.

A statement is thus both a proposition and its local reason, but this reason may be left implicit, requiring then the use of the exclamation mark.

Adding concessions

In the context of the dialogical conception of CTT we also have statements of the form

$$\mathbf{X} \,!\, \pi(x_{1,} \ldots, x_n)[x_i : A_i]$$

where "π" stands for some statement in which $(x_1, \ldots, x_n)$ ocurs, and where $[x_i : A_i]$ stands for some condition under which the statement $\pi(x_1, \ldots, x_n)$ has been brought forward. Thus, the statement reads:

X states that $\pi(x_1, \ldots, x_n)$ under the condition that the antagonist concedes $x_i : A_i$

We call *required concessions* the statements of the form $[x_i : A_i]$ that condition a claim. When the statement is challenged, the antagonist is accepting, through his own challenge, to bring such concessions forward. The concessions of the thesis, if any, are called *initial concessions*. Initial concessions can include formation statements such as $A : prop$, $B : prop$, for the thesis, $A \supset B : prop$.

Formation rules for local reasons: an informal overview

It is presupposed in standard dialogical systems that the players use well-formed formulas (*wff*). The well formation can be checked at will, but only with the usual meta reasoning by which one checks that the formula does indeed observe the definition of a *wff*. We want to enrich our CTT-based dialogical framework by allowing players themselves to first enquire on the formation of the components of a statement within a play. We thus start with dialogical rules explaining the formation of statements involving logical constants (the formation of *elementary* propositions is governed by the Socratic rule, see the discussion above on material truth). In this way, the well formation of the thesis can be examined by the Opponent before running the actual dialogue: as soon as she challenges it, she is *de facto* accepting the thesis to be well formed (the most obvious case being the challenge of the implication, where she has to state the antecedent and thus explicitly endorse it). The Opponent can ask for the formation of the thesis before launching her first challenge; defending the formation of his thesis might for instance bring the Proponent to state that the thesis is a proposition, provided, say, that *A* *is a* ***set*** is conceded; the Opponent might then concede that *A* *is a* ***set***, but only after the constitution of *A* has been established, though if this were the case, we would be considering the constitution of an elementary statement, which is a material consideration, not a formal one.

These considerations yield the following condensed presentation of the logical constants (plus *falsum*), in which " $\mathfrak{K}$" in " $A\mathfrak{K}B$" expresses a connective, and " $\mathfrak{Q}$" in " $(\mathfrak{Q}x : A)\ B(x)$" expresses a quantifier.

Formation rules, condensed presentation

	Connective	Quantifier	Falsum
Move	**X** $A\mathfrak{K}B$:*prop*	**X** $(\mathfrak{Q}x{:}A)\ B(x)$:*prop*	**X** $\bot$:*prop*
Challenge	**Y** $?_{F\mathfrak{K}1}$ and/or **Y** $?_{F\mathfrak{K}2}$	**Y** $?_{F\mathfrak{Q}1}$ and/or **Y** $?_{F\mathfrak{Q}2}$	—
Defence	**X** *A*:*prop* (resp.) **X** *B*:*prop*	**X** *A*:*set* (resp.) **X** $B(x)$:*prop* $(x{:}A)$	—

Synthesis of local reasons

The synthesis rules of local reasons determine how to produce a local reason for a statement; they include rules of interaction indicating how to produce the local reason that is required by the proposition (or set) in play, that is, they indicate what kind of dialogical action—what kind of move—must be carried out, by whom (challenger or defender), and what reason must be brought forward.

Implication. For instance, the synthesis rule of a local reason for the implication $A \supset B$ stated by player **X** indicates:

i. that the challenger **Y** must state the antecedent (while providing a local reason for it): **Y** $p_1 : A$.[9]
ii. that the defender **X** must respond to the challenge by stating the consequent (with its corresponding local reason): **X** $p_2 : B$.

[9]This notation is a variant of the one used by Keiff (2004, 2009).

In other words, the rules for the synthesis of a local reason for implication are as follows:

Synthesis of a local reason for implication

	Implication
Move	**X** ! $A \supset B$
Challenge	**Y** p_1:A
Defence	**X** p_2:B

Notice that the initial statement (**X** ! $A \supset B$) *does not* display a local reason for the claim the the implication holds: player **X** simply states that he has some reason supporting the claim. We express such kind of move by adding an *exclamation mark* before the proposition. The further dialogical actions indicate the moves required for producing a local reason in defence of the initial claim.

Conjunction. The synthesis rule for the conjunction is straightforward:

Synthesis of a local reason for conjunction

	Conjunction
Move	**X** ! $A \wedge B$
Challenge	**Y** ? $L^{\wedge}$ or **Y** ? $R^{\wedge}$
Defence	**X** p_1:A (resp.) **X** p_2:B

Disjunction. For disjunction, as we know from the standard rules, it is the defender who will choose which side he wishes to defend: the challenge consists in requesting of the defender that he chooses which side he will be defending. The point is that each choice is sufficient for defending the claim on the disjunction:

Synthesis of a local reason for disjunction

	Disjunction
Move	**X** ! $A \vee B$
Challenge	**Y** $?_{\vee}$
Defence	**X** p_1:A or **X** p_2:B

The general structure for the synthesis of local reasons

More generally, the rules for the synthesis of a local reason for a constant $\mathcal{K}$is determined by the following triplet:

General structure for the synthesis of a local reason for a constant

	A constant $\mathcal{K}$	Implication	Conjunction	Disjunction
Move	X ! $\varphi[\mathcal{K}]$ *X claims that* ϕ	X ! $A \supset B$	X ! $A \wedge B$	X ! $A \vee B$
Challenge	*Y asks for the reason backing such a claim*	Y p_1:A	Y ? $L^{\wedge}$ or Y ? $R^{\wedge}$	Y $?_{\vee}$
Defence	X p:$\varphi[\mathcal{K}]$ *X states the local reason* p *for* $\varphi[\mathcal{K}]$ *according to the rules for the synthesis of local reasons prescribed for* $\mathcal{K}$	X p_2:B	X p_1:A (resp.) X p_2:B	X p_1:A or X p_2:B

Analysis of local reasons

Apart from the rules for the synthesis of local reasons, we need rules that indicate how to parse a complex local reason into its elements: this is the *analysis* of local reasons. In order to deal with the complexity of these local reasons and formulate general rules for the analysis of local reasons (at the play level), we introduce certain operators that we call *instructions*, such as $L^{\vee}(p)$ or $R^{\wedge}(p)$.

Approaching the analysis rules for local reasons

Let us introduce these instructions and the analysis of local reasons with an example: player **X** states the implication $(A \wedge B) \supset A$. According to the rule for the synthesis of local reasons for an implication, we obtain the following:

Move	X ! $(A \wedge B) \supset B$
Challenge	Y p_1:$A \wedge B$

Recall that the synthesis rule prescribes that **X** must now provide a local reason for the consequent; but instead of defending his implication (with **X** $p_2 : B$ for instance), **X** can choose to parse the reason p_1 provided by **Y** in order to force **Y** to provide a local reason for the right-hand side of the conjunction that **X** will then be able to copy; in other words, **X** can force **Y** to provide the local reason for B out of the local reason p_1 for the antecedent $A \wedge B$ of the initial implication. The analysis rules prescribe how to carry out such a parsing of the statement by using *instructions*. The rule for the analysis of a local reason for the conjunction $p_1 : A \wedge B$ will thus indicate that its defence includes expressions such as

- the left instruction for the conjunction, written $L\wedge(p_1)$, and
- the right instruction for the conjunction, written $R\wedge(p_1)$.

These instructions can be informally understood as carrying out the following step: for the defence of the conjunction $p_1 : A \wedge B$ separate the local reason p_1 on its left (or right) component so that this component can be adduced in defence of the left (or right) side of the conjunction.

Here is a play with local reasons for the thesis $(A \wedge B) \supset B$ using instructions:

	O			**P**	
				$!\ (A \wedge B) \supset B$	0
1	$m := 1$			$n := 2$	2
3	$p_1{:}A \wedge B$	0		$R^{\wedge}(p_1){:}B$	6
5	$R^{\wedge}(p_1){:}B$		3	$?R^{\wedge}$	4

P wins.

In this play, **P** uses the analysis of local reasons for conjunction in order to force **O** to state $R^{\wedge}(p_1) : B$, that is to provide a local reason[10] for the elementary statement B; **P** can then copy that local reason in order to back his statement B, the consequent of his initial implication. With these local reasons, we explicitly have in the object-language the reasons that are given and asked for and which constitute the essence of an argumentative dialogue.

The general structure for the analysis rules of local reasons

	Move	Challenge	Defence
Conjunction	**X** $p{:}A \wedge B$	**Y** $?L^{\wedge}$ or **Y** $?R^{\wedge}$	**X** $L^{\wedge}(p){:}A$ (resp.) **X** $R^{\wedge}(p){:}B$
Disjunction	**X** $p{:}A \vee B$	**Y** $?_{\vee}$	**X** $L^{\vee}(p){:}A$ or **X** $R^{\vee}(p){:}B$
Implication	**X** $p{:}A \supset B$	**Y** $L^{\supset}(p){:}A$	**X** $R^{\supset}(p){:}B$

[10] Speaking of local reasons is a little premature at this stage, since only instructions are provided and not actual local reasons; but the purpose is here to give the general idea of local reasons, and instructions are meant to be resolved into proper local reasons, which requires only an extra step.

Interaction procedures embedded in instructions

Carrying out the prescriptions indicated by instructions require the following three interaction-procedures:

1. *Resolution of instructions*: this procedure determines how to carry out the instructions prescribed by the rules of analysis and thus provide an actual local reason.
2. *Substitution of instructions*: this procedure ensures the following; once a given instruction has been carried out through the choice of a local reason, say *b*, then every time the same instruction occurs, it will always be substituted by the same local reason *b*.
3. *Application of the Socratic rule*: the Socratic rule prescribes how to constitute equalities out of the resolution and substitution of instructions, linking synthesis and analysis together.

Let us discuss how these rules interact and how they lead to the main thesis of this study, namely that immanent reasoning is equality in action.

From Reasons to Equality: a new visit to endorsement

One of the most salient features of dialogical logic is the so-called, *Socratic rule* (or Copy-cat rule or rule for *the formal use of elementary propositions* in the standard—that is, non-CTT—context), establishing that the Proponent can play an elementary proposition only if the Opponent has played it previously.

The Socratic rule is a characteristic feature of the dialogical approach: other game-based approaches do not have it and it relates to *endorsing* condition mentioned in the introduction. With this rule the dialogical framework comes with an internal account of elementary propositions: an account in terms of interaction only, without depending on metalogical meaning explanations for the non-logical vocabulary.

The rule has a clear Platonist and Aristotelian origin and sets the terms for what it is to carry out a *formal argument*: see for instance Plato's *Gorgias* (472b–c). We can sum up the underlying idea with the following statement:

> There is no better grounding of an assertion within an argument than indicating that it has been already conceded by the Opponent or that it follows from these concessions.[11]

What should be stressed here are the following two points:

1. formality is understood as a kind of *interaction*; and
2. formal reasoning *should not* be understood here as devoid of content and reduced to purely syntactic moves.

Both points are important in order to understand the criticism often raised against formal reasoning in general, and in logic in particular. It is only quite late in the history of philosophy that formal reasoning has been reduced to syntactic manipulation— presumably the first explicit occurrence of the syntactic view of logic is

[11] Recent researches on deploying the dialogical framework for the study of history of logic claim that this rule is central to the interpretation of dialectic as the core of Aristotle's logic—see Crubellier (2014, pp. 11–40) and Marion and Rückert (2015).

Leibniz's "pensée aveugle" (though Leibniz's notion was not a reductive one). Plato and Aristotle's notion of formal reasoning is neither "static" nor "empty of meaning". In the Ancient Greek tradition logic emerged from an approach of assertions in which meaning and justification result from what has been brought forward during argumentative interaction. According to this view, dialogical interaction is constitutive of meaning.

Some former interpretations of standard dialogical logic did understand formal plays in a purely syntactic manner. The reason for this is that the standard version of the framework is not equipped to express meaning at the object-language level: there is no way of asking and giving reasons for elementary propositions. As a consequence, the standard formulation simply relies on a syntactic understanding of *Copy-cat moves*, that is, moves entitling **P** to copy the elementary propositions brought forward by **O**, regardless of its content.

The dialogical approach to CTT (dialogues for immanent reasoning) however provides a fine-grain study of the contentual aspects involved in formal plays, much finer than the one provided by the standard dialogical framework. In dialogues for immanent reasoning which we are now presenting, a statement is constituted both by a proposition and by the (local) reason brought forward in defence of the claim that the proposition holds. In formal plays not only is the Proponent allowed to copy an elementary proposition stated by the Opponent, as in the standard framework, but he is also allowed to adduce in defence of that proposition the *same* local reason brought forward by the Opponent when she defended that same proposition. Thus immanent reasoning and equality in action are intimately linked. In other words, a formal play displays the *roots of the content* of an elementary proposition, and *not* a syntactic manipulation of that proposition.

Statements of definitional equality emerge precisely at this point. In particular reflexivity statements such as

$$p = p \,.\, A$$

express from the dialogical point of view the fact that if **O** states the elementary proposition A, then **P** can do the same, that is, play the same move and do it on the same grounds which provide the meaning and justification of A, namely p.

These remarks provide an insight only on simple forms of equality and barely touch upon the finer-grain distinctions discussed above; we will be moving to these by means of a concrete example in which we show, rather informally, how the combination of the processes of analysis, synthesis, and resolution of instructions lead to equality statements.

Example

Assume that the Proponent brings forward the thesis $(A \wedge B) \supset (B \wedge A)$:

	O			P	
				$!\ (A \wedge B) \supset (B \wedge A)$	0

Both players then choose their repetition ranks:

	O			P	
				$!\ (A \wedge B) \supset (B \wedge A)$	0
1	$m := 1$			$n := 2$	2

O must now challenge the implication if she accepts to enter into the discussion. The rule for the synthesis of a local reason for implication (provided above) stipulates that in order to challenge the thesis, **O** must state the antecedent *and provide a local reason for it*:

		O			P	
					$!\ (A \wedge B) \supset (B \wedge A)$	0
	1	$m := 1$			$n := 2$	2
Synthesis of a local reason for conjunction	3	$p{:}A \wedge B$	0			

According to the same synthesis-rule **P** must now state the consequent, which he is allowed to do because the consequent is not elementary:

	O			P	
				$!\ (A \wedge B) \supset (B \wedge A)$	0
1	$m := 1$			$n := 2$	2
3	$p{:}A \wedge B$	0		$q{:}B \wedge A$	4

The Opponent launches her challenge asking for the left component of the local reason q provided by **P**, an application of the rule for the *analysis* of a local reason for a conjunction described above.

Analysis of a local reason for conjunction

	O			**P**	
				$!\ (A \wedge B) \supset (B \wedge A)$	0
1	$m := 1$			$n := 2$	2
3	$p{:}A \wedge B$	0		$q{:}B \wedge A$	4
5	$?\ L^{\wedge}$	4			

Assume that **P** responds immediately to this challenge:

	O			**P**	
				$!\ (A \wedge B) \supset (B \wedge A)$	0
1	$m := 1$			$n := 2$	2
3	$p{:}A \wedge B$	0		$q{:}B \wedge A$	4
5	$?\ L^{\wedge}$	4		$L^{\wedge}(q){:}B$	6

O will now ask for the *resolution of the instruction*:

Resolution of an instruction

	O			**P**	
				$!\ (A \wedge B) \supset (B \wedge A)$	0
1	$m := 1$			$n := 2$	2
3	$p{:}A \wedge B$	0		$q{:}B \wedge A$	4
5	$?\ L^{\wedge}$	4		$L^{\wedge}(q){:}B$	6
7	$?\ \ldots/L^{\wedge}(q)$	6			

In this move 7, **O** is asking **P** to carry out the instruction $L^{\wedge}(q)$ by bringing forward the local reason of his choice. The act of choosing such a reason and replacing the instruction for it is called *resolving the instruction.*

In this case, resolving the instruction will lead **P** to bring forward an *elementary statement*—that is, a statement in which *both* the local reason and the proposition are elementary, which falls under the restriction of the Socratic rule. The idea for **P** then is to postpone his answer to the challenge launched with move 7 and to force **O** to choose a local reason first so as to copy it in his answer to the challenge. This yields a further application of the *analysis rule* for the conjunction:

Comment		O			P		Comment
					$!\,(A \wedge B) \supset (B \wedge A)$	0	
	1	$m := 1$			$n := 2$	2	
	3	$p{:}A \wedge B$	0		$q{:}B \wedge A$	4	
	5	$?\,L^{\wedge}$	4		$L^{\wedge}(q){:}B$	6	
	7	$?\ldots/L^{\wedge}(q)$	6		$b : B$	12	
O responds according to the *analysis rule*	9	$R^{\wedge}(p){:}B$		3	$?\,R^{\wedge}$	8	**P** launches his challenge asking for the right side of the concession move 3
O responds to the challenge by choosing the local reason *b*	11	$b{:}B$		9	$?\ldots/\,R^{\wedge}(p)$	10	**P** asks **O** to resolve the instruction by providing a local reason

P wins.

Move 11 thus provides **P** with the information he needed: he can then copy **O**'s choice to answer the challenge she launched at move 7.

Note: It should be clear that a similar end will come about if **O** starts by challenging the right component of the conjunction statement, instead of challenging the left component.

Analysis of the example

Let us now go deeper in the analysis of the example and make explicit what happened during the play:

When **O** resolves $R^{\wedge}(p)$ with the local reason b (for instance) and **P** resolves the instruction $L^{\wedge}(q)$ with the same local reason, then **P** is not only stating $b : B$ but he is doing this by choosing b as local reason for B, that is, by choosing *exactly the same* local reason as **O** for the resolution of $R^{\wedge}(p)$.

Let us assume that **O** can ask **P** to make his choice for a given local reason explicit. **P** would then answer that his choice for his local reason depends on **O**'s own choice: he simply copied what **O** considered to be a local reason for B, that is $R^{\wedge}(p)^{O} = b : B$. The application of the Socratic rule yields in this respect definitional equality. This rule prescribes the following response to a challenge on an elementary local reason:

When **O** challenges an elementary statement of **P** such as $b : B$, **P** must be able to bring forward a definitional equality such as **P** $R^{\wedge}(p) = b : B$.

Which reads:

P grounds his choice of the local reason b for the proposition B in **O**'s resolution of the instruction $R^{\wedge}(p)$. At the very end **P**'s choice is the *same local reason* brought forward by **O** for the same proposition B.

In other words, the definitional equality $R^{\wedge}(p)^{O} = b : B$ that provides content to B makes it explicit at the object-language level that an application of the Socratic rule has been initiated and achieved by means of dialogical interaction.

The development of a dialogue determined by immanent reasoning thus includes four distinct stages:

1. applying the rules of synthesis to the thesis;
2. applying the rules of analysis;
3. launching the Resolution and Substitution of instructions;
4. applying the Socratic rule.
5. We can then add a fifth stage: Producing the strategic reason.

While the first two steps involve local meaning, step 3 concerns global meaning and step 4 requires describing how to produce a winning strategy. Now that the general idea of local reasons has been provided, we will present in the next chapter all the rules together, according to their level of meaning.

2.2.2 The Dialogical Roots of Equality: Dialogues for Immanent Reasoning

In this section we will spell out a *simplified version* of the dialogues for immanent reasoning, that is, the dialogical framework incorporating features of Constructive Type Theory—a dialogical framework making the players' reasons for asserting a proposition explicit. The rules can be divided, just as in the standard framework, into rules determining local meaning and rules determining global meaning. These include:

1. Concerning *local meaning*
   a. formation rules;
   b. rules for the synthesis of local reasons; and
   c. rules for the analysis of local reasons.
2. Concerning *global meaning*, we have the following (structural) rules:
   a. rules for the resolution of instructions;
   b. rules for the substitution of instructions;
   c. equality rules determined by the application of the Socratic rules.

We will be presenting these rules in this order in the next two sections, along with the adaptation of the other structural rules to dialogues for immanent reasoning in the second section.

2.2.2.1 Local Meaning in Dialogues for Immanent Reasoning

The formation rules

The formation rules for *logical constants* and for *falsum* are given in the following table. Notice that a statement '⊥: **prop**' cannot be challenged; this is the dialogical account for falsum '⊥' being by definition a proposition.

Formation rules

	Move	Challenge	Defence
Conjunction	**X** $A \wedge B$:***prop***	**Y** ? $F_{\wedge 1}$ or **Y** ? $F_{\wedge 2}$	**X** A:***prop*** (resp.) **X** B:***prop***
Disjunction	**X** $A \vee B$:***prop***	**Y** ? $F_{\vee 1}$ or **Y** ? $F_{\vee 2}$	**X** A:***prop*** (resp.) **X** B:***prop***
Implication	**X** $A \supset B$:***prop***	**Y** ? $F_{\supset 1}$ or **Y** ? $F_{\supset 2}$	**X** A:***prop*** (resp.) **X** B:***prop***
Universal quantification	**X** $(\forall x{:}A)B\ x)$:***prop***	**Y** ? $F_{\forall 1}$ or **Y** ? $F_{\forall 2}$	**X** A:***set*** (resp.) **X** $B(x)$:***prop***$[x{:}A]$
Existential quantification	**X** $(\exists x{:}A)B(x)$:***prop***	**Y** ? $F_{\exists 1}$ or **Y** ? $F_{\exists 2}$	**X** A:***set*** (resp.) **X** $B(x)$:***prop***$[x{:}A]$
Falsum	**X** ⊥:***prop***	—	—

The substitution rule within dependent statements

The following rule is not really a formation-rule but it is the dialogical rendering of hypotheticals.[12]

[12]This rule is an expression at the level of plays of the rule for the substitution of variables in a hypothetical judgement. See Martin-Löf (1984, pp. 9–11).

Hypotheticals

Substitution rule within dependent statements (subst-D)

	Move	Challenge	Defence
Subst-D	**X** $B[x_1, \ldots, x_n]$ $[x_1{:}A_1, x_2{:}A_2[x_1], \ldots, x_n{:}A_n[x_1, \ldots, x_{n-1}]]$	**Y** $p_1{:}A_1, \ldots, p_n{:}A_n$	**X** $B(p_1, \ldots, p_n)$

A particular case of the application of Subst-D is when the challenger simply chooses the same local reasons as those occurring in the concession of the initial statement. This is particularly useful in the case of formation plays:

The rules for local reasons: synthesis and analysis

Now that the dialogical account of formation rules has been clarified, we may further develop our analysis of plays by introducing local reasons. Let us do so by providing the rules that prescribe the synthesis and analysis of local reasons.

Synthesis rules for local reasons

	Move	Challenge	Defence
Conjunction	**X** ! $A \wedge B$	**Y** ? $L^{\wedge}$ or **Y** ? $R^{\wedge}$	**X** $p_1{:}A$ (resp.) **X** $p_2{:}B$
Existential quantification	**X** ! $(\exists x{:}A)B(x)$	**Y** ? $L^{\exists}$ or **Y** ? $R^{\exists}$	**X** $p_1{:}A$ (resp.) **X** $p_2{:}B(p_1)$
Disjunction	**X** ! $A \vee B$	**Y** ?$^{\vee}$	**X** $p_1{:}A$ or **X** $p_2{:}B$
Implication	**X** ! $A \supset B$	**Y** $p_1{:}A$	**X** $p_2{:}B$
Universal quantification	**X** ! $(\forall x{:}A)B(x)$	**Y** $p_1{:}A$	**X** $p_2{:}B(p_1)$
Negation	**X** ! $\neg A$ Also expressed as **X** ! $A \supset \bot$	**Y** $p_1{:}A$	**X** $p_2{:}\bot$

Analysis rules for local reasons

	Move	Challenge	Defence
Conjunction	**X** $p{:}A \wedge B$	**Y** ? $L^{\wedge}$ or **Y** ? $R^{\wedge}$	**X** $L^{\wedge}(p){:}A$ (resp.) **X** $R^{\wedge}(p){:}B$
Existential quantification	**X** $p{:}(\exists x{:}A)B(x)$	**Y** ? $L^{\exists}$ or **Y** ? $R^{\exists}$	**X** $L^{\exists}(p){:}A$ (resp.) **X** $R^{\exists}(p){:}B(L^{\exists}(p))$
Disjunction	**X** $p{:}A \vee B$	**Y** $?^{\vee}$	**X** $L^{\vee}(p){:}A$ or **X** $R^{\vee}(p){:}B$
Implication	**X** $p{:}A \supset B$	**Y** $L^{\supset}(p){:}A$	**X** $R^{\supset}(p){:}B$
Universal quantification	**X** $p{:}(\forall x{:}A)B(x)$	**Y** $L^{\forall}(p){:}A$	**X** $R^{\forall}(p){:}B(L^{\forall}(p))$
Negation	**X** $p{:}\neg A$ Also expressed as **X** $p{:}A \supset \bot$	**Y** $L^{\neg}(p){:}A$ **Y** $L^{\supset}(p){:}A$	**X** $R^{\neg}(p){:}\bot$ **X** $R^{\supset}(p){:}\bot$ Which amounts to stating **X** ! ⊥ [a]

[a]The general point of deleting the instruction in **X** $R^{\supset}(p){:}\bot$ is that instructions occurring in expressions stating **falsum** keep un-resolved – see below structural rule SR3 on resolutions, item 3

Anaphoric instructions: dealing with cases of anaphora

One of the most salient features of the CTT framework is that it contains the means to deal with cases of anaphora. For example anaphoric expressions are required for formalizing *Barbara* in CTT. In the following CTT-formalization of *Barbara* the projection fst(z) can be seen as the tail of the anaphora whose head is z:

$$(\forall z : (\exists x : D)A)B[\text{fst}\ (z)]\ \text{true} \quad \text{premise 1}$$
$$(\forall z : (\exists x : D)B)C[\text{fst}(z)]\ \text{true} \quad \text{premise 2}$$
$$\overline{\qquad\qquad\qquad\qquad\qquad\qquad}$$
$$(\forall z : (\exists x : D)A)C[\text{fst}(z)]\ \text{true} \quad \text{conclusion}$$

In dialogues for immanent reasoning, when a local reason has been made explicit, this kind of anaphoric expression is formalized through instructions, which provides a further reason for introducing them. For example if p is the local reason for the first premise we have

$$\mathbf{P}p : (\forall z : (\exists x : D)A(x))B\big(L^{\exists}\big(L^{\forall}(p)^{\mathbf{O}}\big)\big)$$

However, since the thesis of a play does not bear an explicit local reason (we use the exclamation mark to indicate there is an implicit one), it is possible for a statement to be bereft of an explicit local reason. When there is no explicit local reason for a statement using anaphora, we cannot bind the instruction $L^{\forall}(p)^{\mathbf{O}}$ to a local reason p. We thus have something like this, with a blank space instead of the anaphoric local reason:

$$\mathbf{P}\,!\,(\forall z : (\exists x : D)A(x))B\big(L^{\exists}\big(L^{\forall}(\ )^{\mathbf{O}}\big)\big)$$

But this blank stage can be circumvented: the challenge on the universal quantifier will yield the required local reason: O will provide $a : \big(\exists x : D\big)A(x)$, which is the local reason for z. We can therefore bind the instruction on the missing local reason with the corresponding variable—z in this case—and write

$$\mathbf{P}\,!\,(\forall z : (\exists x : D)A(x))B\big(L^{\exists}\big(L^{\forall}(z)^{\mathbf{O}}\big)\big)$$

We call this kind of instruction, Anaphoric instructions. For the substitution of Anaphoric instructions the following two cases are to be distinguished:

Substitution of Anaphoric Instructions 1

Given some Anaphoric instruction such as $L^{\forall}(z)^{\mathbf{Y}}$, once the quantifier $\big(\forall z : A\big)B(\ldots)$ has been challenged by the statement $a : A$, the occurrence of $L^{\forall}(z)^{\mathbf{Y}}$ can be substituted by a. The same applies to other instructions.

In our example we obtain:

$$\begin{array}{c} \mathbf{P}\,!\,(\forall z : (\exists x : D)A(x))B\big(L^{\exists}\big(L^{\forall}(z)^{\mathbf{O}}\big)\big) \\ \mathbf{O}a : (\exists x : D)A(x) \\ \mathbf{P}b : B\big(L^{\exists}\big(L^{\forall}(z)^{\mathbf{O}}\big)\big) \\ \mathbf{O}\,?\,a/L^{\forall}(z)^{\mathbf{O}} \\ \mathbf{P}b : B\big(L^{\exists}(a)\big) \\ \ldots \end{array}$$

Substitution of Anaphoric Instructions 2

Given some Anaphoric instruction such as $L^{\forall}(z)^{\mathbf{Y}}$, once the instruction $L^{\forall}(c)$—resulting from an attack on the universal $\forall z : \varphi$ — has been resolved with $a : \varphi$, then any occurrence of $L^{\forall}(z)^{\mathbf{Y}}$ can be substituted by a. The same applies to other instructions.

2.2.2.2 Global Meaning in Dialogues for Immanent Reasoning

We here provide the structural rules for dialogues for immanent reasoning, which determine the global meaning in such a framework. They are for the most part similar in principle to the precedent logical framework for dialogues; the rules concerning instructions are an addition for dialogues for immanent reasoning.

Structural Rules

SR0: Starting rule

The start of a formal dialogue of immanent reasoning is a move where **P** states the thesis. The thesis can be stated under the condition that **O** commits herself to certain other statements called initial concessions; in this case the thesis has the form ! A [B_1, ..., B_n], where A is a statement with implicit local reason and $B_1, \ldots, B_n$ are statements with or without implicit local reasons.
A dialogue with a thesis proposed under some conditions starts if and only if **O** accepts these conditions. **O** accepts the conditions by stating the initial concessions in moves numbered 0.1, ..., 0.n before choosing the repetition ranks.
After having stated the thesis (and the initial concessions, if any), each player chooses in turn a positive integer called the repetition rank which determines the upper boundary for the number of attacks and of defences each player can make in reaction to each move during the play.

SR1: Development rule

The Development rule depends on what kind of logic is chosen: if the game uses intuitionistic logic, then it is SR1i that should be used; but if classical logic is used, then SR1c must be used.

SR1i: Intuitionistic Development rule, or Last Duty First

Players play one move alternately. Any move after the choice of repetition ranks is either an attack or a defence according to the rules of formation, of synthesis, and of analysis, and in accordance with the rest of the structural rules. Players can answer only against the *last non-answered* challenge by the adversary.

Note: This structural rule is known as the Last Duty First condition, and makes dialogical games suitable for intuitionistic logic, hence the name of this rule.

SR1c: Classical Development rule

Players play one move alternately. Any move after the choice of repetition ranks is either an attack or a defence according to the rules of formation, of synthesis, and of analysis, and in accordance with the rest of the structural rules.
If the logical constant occurring in the thesis is not recorded by the table for local meaning, then either it must be introduced by a nominal definition, or the table for local meaning needs to be enriched with the new expression.

Note: The structural rules with SR1c (and not SR1i) produce strategies for classical logic. The point is that since players can answer to a list of challenges in any order (which is not the case with the intuitionistic rule), it might happen that the two options of a **P**-defence occur in the same play—this is closely related to the classical development rule in sequent calculus allowing more than one formula at the right of the sequent.

SR2: Formation rules for formal dialogues
SR2i: Starting a formation dialogue

A formation-play starts by challenging the thesis with the formation request **O** $?_{\mathbf{prop}}$; **P** must answer by stating that his thesis is a proposition.

SR2ii: Developing a formation dialogue

The game then proceeds by applying the formation rules up to the elementary constituents of **prop/set**.
After that **O** is free to use the other particle rules insofar as the other structural rules allow it.

Note: The constituents of the thesis will therefore not be specified before the play but as a result of the structure of the moves (according to the rules recorded by the rules for local meaning).

SR3: Resolution of instructions

1. A player may ask his adversary to carry out the prescribed instruction and thus bring forward a suitable local reason in defence of the proposition at stake. Once the defender has replaced the instruction with the required local reason we say that the instruction has been resolved.
2. The player index of an instruction determines which of the two players has the right to choose the local reason that will resolve the instruction.
   a. If the instruction $\mathfrak{I}$ for the logical constant $\mathcal{K}$ has the form $\mathfrak{I}^{\mathcal{K}}(p)^{\mathbf{X}}$ and it is **Y** who requests the resolution, then the request has the form **Y** ?.../ $\mathfrak{I}^{\mathcal{K}}(p)^{\mathbf{X}}$, and it is **X** who chooses the local reason.
   b. If the instruction $\mathfrak{I}$ for the logic constant $\mathcal{K}$ has the form $\mathfrak{I}^{\mathcal{K}}(p)^{\mathbf{Y}}$ and it is player **Y** who requests the resolution, then the request has the form **Y** p_i / $\mathfrak{I}^{\mathcal{K}}(p)^{\mathbf{Y}}$, and it is **Y** who chooses the local reason.
3. Instructions occurring in expressions stating **falsum** have no resolution. In fact, The player stating $\mathfrak{I}(p){:}\bot$ gives up and therefore loses the play.

SR4: Substitution of instructions

Once the local reason b has been used to resolve the instruction $\mathfrak{I}^{\mathcal{K}}(p)^{\mathbf{X}}$, and if the same instruction occurs again, players have the right to require that the instruction be resolved with b. The substitution request has the form $?\boldsymbol{b}/\mathfrak{I}_k(p)^{\mathbf{X}}$. Players cannot choose a different substitution term (in our example, not even **X**, once the instruction has been resolved).
This rule also applies to functions.

SR5: Socratic rule and definitional equality

The following points are all parts of the Socratic rule, they all apply.

SR5.1: Restriction of P statements

P cannot make an elementary statement if **O** has not stated it before, except in the thesis.
An elementary statement is either an elementary proposition with implicit local reason, or an elementary proposition and its local reason (not an instruction).

SR5.2: Challenging elementary statements in formal dialogues

Challenges against elementary statements with implicit local reasons take the form:

$\mathbf{X}\ !A$
$Y?_{reason}$
$\mathbf{X}\ a : A$

where A is an elementary proposition and a is a local reason.[13]
P cannot challenge **O**'s elementary statements, except if **O** provides an elementary initial concession with implicit local reason, in which case **P** can ask for a local reason, or in the context of transmission of equality.

SR5.3: Definitional equality

O may challenge elementary **P**-statements; **P** then answers by stating a definitional equality expressing the equality between a local reason and an instruction both introduced by **O** (for non-reflexive cases, that is when **O** provided the local reason as a resolution of an instruction), or a reflexive equality of the local reason introduced by **O** (when the local reason was not introduced by the resolution of an instruction, that is either as such in the initial concessions or as the result of a synthesis of a local reason). We thus distinguish two cases of the Socratic rule:

1. non-reflexive cases;
2. reflexive cases.
   These rules do not cover cases of transmission of equality. The Socratic rule also applies to the resolution or substitution of functions, even if the formulation mentions only instructions.

SR5.3.1: Non-reflexive cases of the Socratic rule

We are in the presence of a *non-reflexive case* of the Socratic rule when **P** responds to the challenge with the indication that **O** gave the same local reason for the same proposition when she had to resolve or substitute instruction I.

---

[13]Note that **P** is allowed to make an elementary statement only as a thesis (Socratic rule); he will be able to respond to the challenge on an elementary statement only if **O** has provided the required local reason in her initial concessions.

Here are the different challenges and defences determining the meaning of the three following moves:

<u>**Non-reflexive cases of the Socratic rule**</u>

	Move	Challenge	Defence
SR5.3.1a	**P** $a{:}A$	**O** $?{=}\,a$	**P** $I = a{:}A$
SR5.3.1b	**P** $a{:}A(b)$	**O** $?{=}\,b^{A(b)}$	**P** $I = b{:}D$
SR5.3.1c	**P** $I = b{:}D$ (this statement stems from **SR5.3.1b**)	**O** $? = A(b)$	**P** $A(I) = A(b){:}\textbf{\textit{prop}}$

Presuppositions:

(i) The response prescribed by SR5.3.1a presupposes that **O** has stated A or $a = b : A$ as the result of the resolution or substitution of instruction I occurring in I : A or in I $= b : A$.

(ii) The response prescribed by SR5.3.1b presupposes that **O** has stated A and $b : D$ as the result of the resolution or substitution of instruction I occurring in $a : A(I)$.

(iii) SR5.3.1c assumes that **P** $I = b : D$ is the result of the application of SR5.3.1b. The further challenge seeks to verify that the replacement of the instruction produces an equality in **prop**, that is, that the replacement of the instruction with a local reason yields an equal proposition to the one in which the instruction was not yet replaced. The answer prescribed by this rule presupposes that **O** has already stated $A(b)$: **prop** (or more trivially $A(I) = A(b)$: **prop**).

(iv) The **P**-statements obtained after defending elementary **P**-statements cannot be attacked again with the Socratic rule (with the exception of SR5.3.1c), nor with a rule of resolution or substitution of instructions.

SR5.3.2: Reflexive cases of the Socratic rule

We are in the presence of a *reflexive case* of the Socratic rule when **P** responds to the challenge with the indication that **O** adduced the same local reason for the same proposition, though that local reason in the statement of **O** is not the result of any resolution or substitution.

Attacks have the same form as those prescribed by SR5.3.1. Responses that yield reflexivity presuppose that **O** has previously stated the same statement or even the same equality.

The response obtained cannot be attacked again with the Socratic rule.

SR6: Transmission of Definitional Equality

Definitional Equality transmits by reflexivity, transitivity and symmetry.

SR7: Winning rule for plays

The player who makes the last move wins unless he states $\bot$. The player who states **falsum** loses the play.[14]

Resolution and the Justification of the Analysis Rules for Local Reasoning

Notice that the analysis rules for local reasons meet the justification criteria required by constructivist theories of meaning; but at the play level.

Indeed, according to the constructivist approach in order to justify an elimination rule it is necessary to make the conclusion evident, on the assumption that the premises of the rule are known. In the CTT framework this is achieved by showing that, if $d : C$ is the conclusion of an elimination rule, then d evaluates to a canonical element of C (i.e. d evaluates to an element occurring in the conclusion of an introduction rule for C). The procedure of evaluation consists in the unwinding of definitions (implemented by suitable equality rules), replacing defined expressions by their definientia.

In the dialogical setting justifying a rule of analysis at the play level for some claim **X** $d : C$ amounts to the task of showing that the "resolution" (or execution) of the instruction(s) prescribed by that analysis-rule render those local reasons determined by the synthesis rule for the claim **X** ! C. Moreover, the justification should show that the analysis rule for **X** $d : C$, does not contravene the distribution of rights and duties associated by the synthesis rule to the claim **X** ! C. An informal argument for the justification of the analysis-rules for local reasons is straightforward.

X$p : A \wedge B$. If player **X** states a conjunction backing it with local reason p, then the corresponding instructions once resolved, render back, what the synthesis rules prescribe, namely that **X** must provide a local reason for the left side, when the antagonist **Y** asks for the left. Similarly for the right side.

[14]See, above point 3 of SR3. At the strategy level the move **O** ! $\bot$ allows **P** to bring forward the strategic reason $you_{gaveup}(n)$ in support for any statement that he has not defended before **O** stated $\bot$ at move n.

$\mathbf{X}p : (\exists x : A)B(x)$ If player **X** states an existential backing it with local reason p, then the corresponding instructions once resolved, render back, what the synthesis rules prescribe, namely that **X** must provide a local reason for the left side, when the antagonist **Y** asks for the left. If the antagonist **Y** asks for the right side, **X** must provide a local reason for the right side (where the local reason for the left side, chosen by **X**, occurs). That is, if **X** $L^{\exists}(p) = p_1 : A$, then **X** $R^{\exists}(p) = p_2 : B(p_1)$.
$\mathbf{X}p : A \vee B$. If player **X** states a disjunction backing it with local reason p, then the corresponding instructions once resolved, render back, what the synthesis rules prescribe, namely that **X** must choose if p provides a local reason either for the left side or the right side.
$\mathbf{X}p : A \supset B$ If player **X** states an implication backing it with local reason p, then the corresponding instructions once resolved, render back, what the synthesis rules prescribe, namely that **X** must provide a local reason for the consequent of the implication, given that the antagonist **Y** provided a local reason for the antecedent.
$\mathbf{X}p : (x : A)B(x)$. If player **X** states a universal backing it with local reason p, then the corresponding instructions once resolved, render back, what the synthesis rules prescribe, namely that **X** must provide a local reason for the right side of the universal, given that the antagonist **Y** provided a local reason for the left side. That is, if **Y** $L^{\forall}(p) = p_1 : A$, then **X** $R^{\forall}(p) = p_2 : B(p_1)$.
The justification of the analysis rule for negation follows the argument for implication. The justification of **falsum** is vacuously satisfied since there is no synthesis rule for it—cf. Sundholm (2012). Moreover, from the dialogical point of view the meaning of **falsum**, since it is an elementary proposition, must actually be considered as a special case of the Socratic Rule (see rule SR7).

2.2.3 *Content and Material Dialogues*

As pointed out by Krabbe (1985, p. 297), *material dialogues*—that is, dialogues in which propositions have content—receive in the writings of Paul Lorenzen and Kuno Lorenz priority over formal dialogues: material dialogues constitute the *locus* where the logical constants are introduced. However in the standard dialogical framework, since both material and formal dialogues marshal a purely syntactic notion of the formal rule—through which logical validity is defined, this contentual feature is bypassed,[15] with this consequence that Krabbe and others after him considered that, after all, *formal* dialogues had priority over material ones.

As can be gathered from the above discussion, we believe that this conclusion stems from shortcomings of the standard framework, in which local reasons are not expressed at the object-language level. We thus explicitly introduced these local reasons in order to undercut this apparent precedence of a formalistic approach that makes away with the contentual origins of the dialogical project.

[15]Krabbe (1985, p. 297).

In fact, in principle, a local reason prefigures a material dialogue displaying the content of the proposition stated. This aspect makes up the ground level of the normative approach to meaning of the dialogical framework, in which *use*—or dialogical interaction—is to be understood as *use prescibed by a rule*; such a use is what Peregrin (2014, pp. 2–3) calls the *role of a linguistic expression*. Dialogical interaction is this *use*, entirely determined by rules that give it meaning: the linguistic expression of every statement determines this statement by the role it plays, that is by the way it is used, and this use is governed by rules of interaction. The meaning of elementary propositions in dialogical interaction thus amounts to their *role* in the kind of interaction that is governed by the Socratic and Global rules for material dialogues, that is by the specific formulations of the Socratic and Global rules for precisely those very propositions.

It follows that material dialogues are important not only for the general issue on the normativity of logic but also for rendering a language with *content*.

We cannot in the present writing fully develop these kind of dialogues, however we will present briefly the case of the set **Bool** which provides the elements to tackle the case of empirical propositions.[16] The latter allows for expressing classical truth-functions within the dialogical framework, and it has an important role in the CTT-approach to empirical propositions.[17] We invite the reader to visit the chapter on material dialogues in Rahman et al. (2018), where discuss material dialogues that include sets of natural numbers and the set **Bool**.

2.2.3.1 The set Bool and Empirical Propositions

Most of the literature differentiating the philosophical perspective underlying the work of Boole and the one of Frege focused on discussing either the different ways both authors understood the relation between logic and psychology or the links between mathematics and logic, or both. According to these studies, Boole's framework has been mainly conceived as a kind of psychologism and a programme for the mathematization of logic, contrasting as well with Frege's radical anti-psychologism as with his logicist project for the foundations of mathematics.

These comparative studies have also been combined with the contrast between model-theoretical approaches to meaning, with their associated notion of *varying domains of discourse*, versus inferentialist approaches to meaning, with a *fixed universe of discourse*. It might be argued that while the first approach could be more naturally understood as an offspring of the algebraic tradition of Boole-Schröder, the second approach could be seen as influenced by Frege's *Begriffsschrift*.[18]

---

[16]Here again we thank to the reviewers who urged us to sketch at least an example of material dialogues.

[17]See Martin-Löf (2014).

[18]Recall the distinction of language *as the universal medium* and *as a calculus* (van Heijenoort 1967).

However, from the point of view of contemporary classical logic, and after the meta-mathematical perspective of Gödel, Bernays, and Tarski, both Boole's and Frege's view on semantics are subsumed under the same formalization, according to which classical semantics amount to a function of interpretation between the sentences ***S*** of a given language ***L*** and the set of truth values $\{\mathbf{0}, \mathbf{1}\}$—let us call such a set the set **Bool**. This function assumes that the well-formed formulae of ***S*** are made dependent upon a domain—either a *local domain of discourse* (in the case of Tarski's-style approach to Boolean-algebra), or a *universal domain* (in the case of Frege). More precisely, this functional approach is based on a separation of cases that simply assumes that the quantifiers and connectives take propositional functions into classical propositions—for a lucid insight on this perspective's limitations see (Sundholm 2006). In fact, the integration of both views within the same formal semantic closes a gap in Boole's framework, which was already pointed out by Frege: the links between the semantics of propositional and first-order logic.[19]

Constructive Type Theory includes a third (epistemic) paradigm in the framework allowing for a new way of dividing the waters between Boolean operators and logical connectives, and, at the same time, integrating them in a common inferential system in which each of them has a specific role to play. The overall paradigm at stake is the Brouwer–Heyting–Kolmogorov conception of propositions as sets of proofs embedded in the framework in which, thanks to the insight brought forward by the Curry–Howard isomorphism (Howard 1980), propositions are read as sets and as types.

In a nutshell, as already mentioned, within CTT the simplest form of a judgement (the categorical) is an expression of the form

$$a : B$$

where "*B*" is a proposition and "*a*" a proof-object on the grounds of which the proposition *B* is asserted to be true, standing as shorthand for

"*a provides evidence for B*".

In other words, the expression "*a* : *B*", is the formal notation of the categorical judgement

"*The proposition B is true*",

which is shorthand for

"*There is evidence for B*".

[19]Frege points out that within Boole's approach there is no organic link between propositional and first-order logic: "*In Boole the two parts run alongside one another; so that one is like the mirror image of the other, but for that very reason stands in no organic relation to it*" (Frege G., Boole's Logical Calculus and the Concept Script [1880/81], 1979).

According to this view, a proposition is a set of elements, called proof-objects, that make the proposition B true. Furthermore, we distinguish between *canonical proof-objects* on the one hand, those entities providing a *direct evidence* for the truth the proposition B, and on the other hand *non-canonical proof-objects*, the entities providing indirect pieces of evidence for B.

This generalization also allows another third reading: a proposition is a *type* and its elements are instances of this type. If we follow this reading proof-objects are conceived as instantiations of the type. This type-reading naturally leads to Brouwer–Heyting–Kolmogorov's constructivism mentioned above: if a proposition is understood as the set of its proofs, it might be the case that we do not have any proof for that proposition at our disposal, but that we neither have a proof for its negation (thus, in such a framework, *third excluded* fails). Notice that the constructivist interpretation requires the intensional rather than the extensional constitution of sets—recall the Aristotelian view that no "form" ("type") can be conceived independently of its instances and reversewise.

Moreover CTT provides a novel way to render the meaning of the set $\{\mathbf{0}, \mathbf{1}\}$ as the type **Bool**. More precisely, the type **Bool** is characterized as the set of the canonical elements **0** and **1**. Thus, each non-canonical element is equal to one of them. But what kind of entities are those (non-canonical) elements that might be equal to **1** or **0**? Since in such a setting **1**, **0** and those equal to them are elements, they are not considered to be of the type proposition; they are rather providers of truth or falsity of a proposition (or a set, according to the Curry–Howard isomorphism between propositions, sets, and types): they are proof-objects that provide evidence for the assertion **Bool true**.

In order to illustrate our point here and to explicit the link with material dialogues, let us take an example outside of mathematics, for instance this *sentence*:

Bachir Diagne is from Senegal.

This sentence differs from the *proposition*, that is, what Frege called the *sense* or *thought* expressed by that sentence, which would be

that Bachir Diagne is from Senegal

So if we take the sentence as expressing the proposition, then we might be able to bring forward some proof-object—some piece of evidence a, his passport or his birth certificate for instance—that renders the proposition *that Bachir Diagne is from Senegal* true. In such a case we have the assertion that the proposition is true on the grounds of the piece of evidence a (the passport), which we can write:

passport : *Bachir Diagne is from Senegal*

or, in a more general assertion:

That Bachir Diagne is from Senegal **true**
(*there is some piece of evidence that Bachir Diagne is from Senegal*)

In this fashion, if we take the sentence *Bachir Diagne is from Senegal* as related to a *Boolean object*, then this sentence triggers a procedure yielding a non-canonical element, say **X** (the proposition), of the set **Bool**. In such a case the sentence would not *express* a proposition, but it could be understood as an *answer* to the question:

Is Bachir Diagne from Senegal?

the answer being:

yes, *Bachir Diagne is from Senegal*

which would thus yield the outcome **1**. In other words, determining to which of the canonical elements, **1** or **0**, the non-canonical element **X** is equal, would require answering to the question *Is Bachir Diagne from Senegal* ? Thus, in our case, we take it to be equal to **1**.[20] The procedure would amount to the following steps:

Is Bachir Diagne from Senegal ?

⇓

yes, *Bachir Diagne is from Senegal*

⇓

X = 1:Bool

The arrows indicate that determining to which of the elements **X** is equal actually is the result of an enquiry (in this case an empirical one). This is not only different from:

passport : *Bachir Diagne is from Senegal*

but it is also different from:

Bool true

[20]For the interpretation of empirical propositions see (Martin-Löf 2014).

Indeed, while $\mathbf{X} = \mathbf{1} : \mathbf{Bool}$ expresses one of the possible outcomes the non-canonical element **X** can take in **Bool**, **Bool true** expresses the fact that at least one element of the set **Bool** can be brought forward.

Thus, a distinction is drawn between the Boolean object **1** (one of the canonical elements of **Bool**) and the predicate **true** that applies to **Bool**.

Moreover, operations between elements of **Bool** would then not be the logical connectives introduced by natural deduction rules at the right hand side of the colon, but they would be operations between objects occurring at the left hand side of the colon. For example the disjunction "+" at left of the colon in

$$A + B = \mathbf{1} : \mathbf{Bool}\ (given\, A = \mathbf{1} : \mathbf{Bool})$$

stands for an operation between the non-canonical **Boolean** objects A and B; whereas the disjunction occurring at right of the colon in the assertion

$$b : A \vee B\ (given\, b : A)$$

expresses the familiar logical connective of disjunction, that is here true since a piece of evidence for one of the disjuncts is provided: the piece of evidence b for A.

Since **Bool** is a type, and since according to the Curry–Howard isomorphism, it is itself a proposition, we can certainly have both, propositional connectives as sets of proof-objects, and have them combined with Boolean operations. This allows us, for example, to demonstrate that each canonical element in **Bool** is identical either to **1** or **0**:

$$(\forall x : \mathbf{Bool})Id(\mathbf{Bool}, x, \mathbf{1}) \vee \mathbf{Id}(\mathbf{Bool}, x, \mathbf{0})\mathbf{true}$$

2.2.3.2 Dialogical Rules for Boolean Operators

In the dialogical framework, the elements of **Bool** are responses to **yes-no** questions, so that each element of **Bool** is equal to **yes** or **no**. Responses such as $b = \mathbf{yes}$ or $b = \mathbf{no}$ make explicit one of the possible origins of the answer **yes** (or **no**), namely whether b is or not the case. Here are the Global (player independent) rules for synthesis, analysis, and equalities of the Boolean operators.

Global rules for Bool and Boolean operators in immanent reasoning

	Move	Challenge	Defence
Synthesis	**X ! Bool**	**Y ?** $_{Bool}$	**X** yes**:Bool**
			X no**:Bool**
Analysis	**X** p:$C(c)[c$:**Bool**]	**Y** $?=c^{\mathbf{Bool}}$	**X** c = yes:**Bool**
			X c = no:**Bool**
Equalities yes	**X** $c=$ yes:**Bool** … **X** p:$C(c)[c$:**Bool**]	**Y** $?=_{reason}$ yes	**X** p_1:$C(yes)$
no	**X** $c=$ no:**Bool** … **X** p:$C(c)[c$:**Bool**]	**Y** $?=_{reason}$ no	**X** p_1:$C(no)$

Specific Socratic Rule for Bool and Boolean Operators

When **O** states a : **Bool**, she is stating that a is an element of **Bool**, that is, she is committing to a being either **yes** or **no**; **P** may challenge this **O**-statement by requesting that she makes her commitment explicit and provides the equality $a =$ **yes** : **Bool** or $a =$ **no** : **Bool**. The following table provides the dialogical rule for this interaction; this rule is part of the Socratic rules because it is player dependent but it is a rule specific to **Bool** and the Boolean operators thus providing their specific meaning: this specific Socratic rule for **Bool** and the Boolean operators provides their material meaning.

Specific Socratic rule for Bool

	Move	Challenge	Defence
Specific Socratic rule for Bool	**O** a:**Bool**	**P** $?=a^{\mathbf{Bool}}$	**O** $a=$ yes:**Bool**
			O $a=$ no:**Bool**

We can now introduce quite smoothly the rules for the classical truth-functional connectives as operations between *elements* of **Bool** (left-hand side of the colon), which are distinct from the usual propositional connectives (right-hand side of the colon). We leave the description for quantifiers to the diligence of the reader, whereby the universal quantifier is understood as a finite sequence of products, and, dually, the existential as a finite sequence of additions.

The dialogical interpretation of the rules below is very close to the usual one: it amounts to the commitments and entitlements specified by the rules of the dialogue: if for instance the response is **yes** to a (left-hand side) product, then the speaker is also committed to answer **yes** to further questions on both of the components of that product. Here again, the meaning of the Boolean operators is provided by interaction, where *choice* is a fundamental feature.

Global rules for classical truth-functional operators

	Move	Synthesis of local reasons	
		Challenge	Defence
Product	**X** $a \times b$:**Bool**	**Y** ?= $a \times b$	**X** $(a \times b) =$ ***yes***:**Bool**
			X $(a \times b) =$ ***no***:**Bool**
***Yes*-equality (product)**	**X** $(a \times b) =$ ***yes***:**Bool**	**Y** ? $L^{\times}$***yes***	**X** $a =$ ***yes***:**Bool**
		Y ? $R^{\times}$***yes***	**X** $b =$ ***yes***:**Bool**
***No*-equality (product)**	**X** $(a \times b) =$ ***no***:**Bool**	**Y** ?$^{\times}$ ***no***	**X** $a =$ ***no***:**Bool**
			X $b =$ ***no***:**Bool**
Addition	**X** $a + b$:**Bool**	**Y** ?= $a + b$	**X** $a =$ ***yes***:**Bool**
			X $b =$ ***yes***:**Bool**
***Yes*-equality (addition)**	**X** $(a + b) =$ ***yes***:**Bool**	**Y** ?$^{+}$ ***yes***	**X** $a =$ ***yes***:**Bool**
			X $b =$ ***yes***:**Bool**
***No*-equality (addition)**	**X** $(a + b) =$ ***no***:**Bool**	**Y** ? L^{+}***no***	**X** $a =$ ***no***:**Bool**
		Y ? R^{+}***no***	**X** $b =$ ***no***:**Bool**
Implication	**X** $a \to b$:**Bool**	**Y** ? = $a \to b$	**X**$(a \to b) =$ ***yes***:**Bool**
			X$(a \to b) =$ ***no***:**Bool**
***Yes*-equality (implication)**	**X** $(a \to b) =$ ***yes***:**Bool**	**Y** $a =$ ***yes***:**Bool**	**X** $b =$ ***yes***:**Bool**
		Y $b =$ ***no***:**Bool**	**X** $a =$ ***no***:**Bool**
***No*-equality (implication)**	**X** $(a \to b) =$ ***no***:**Bool**	**Y** ? $L^{\to}$***no***	**X** $a =$ ***yes***:**Bool**
		Y ? $R^{\to}$***no***	**X** $b =$ ***no***:**Bool**
Negation	**X** $\sim a$:**Bool**	**Y** ? $\sim a$	**X** $\sim a =$ ***yes***:**Bool**
			X $\sim a =$ ***no***:**Bool**
***Yes*-equality (negation)**	**X** $\sim a =$ ***yes***:**Bool**	**Y** ?$^{\sim}$ ***yes***	**X** $a =$ ***no***:**Bool**
***No*-equality (negation)**	**X** $\sim a =$ ***no***:**Bool**	**Y** ?$^{\sim}$ ***no***	**X** $a =$ ***yes***:**Bool**

2.2.3.3 Empirical Propositions

As already mentioned above, non-canonical elements of the set **Bool** can be used to study the meaning of empirical propositions, though what we need in particular is the notion of *empirical quantity*. This notion has been introduced by Martin-Löf in applying CTT to the empirical realm (Martin-Löf 2014): whereas quantities of mathematics and logic are determined by computation, empirical quantities are determined by experiments and observation. An example of a mathematical quantity is 2 + 2; it is determined by a computation yielding the number 4. An example of an empirical quantity is the colour of some object. This is not determined by computation; rather, one must look at the object under normal conditions.

In the dialogical framework, we can consider empirical quantities as answers to a question. For example, give the question

Are Cheryl's eyes blue?

The yes or no answer, obtained through some kind of empirical procedure received in a given context, can be defined over the set **Bool**, namely as being equal to ***yes*** or ***no***. The following question however might involve many different answers:

What is the colour of Cheryl's eyes?

If $\boldsymbol{X}$ stands for the empirical quantity *Colour of Cheryl's eyes*, we might define the possible answers over some finite set $\mathbb{N}^n$ of natural numbers:

$$\begin{aligned}
&\boldsymbol{X} = 1 : \mathbb{N}^n \text{ if Cheryl's eyes are brown}\\
&\boldsymbol{X} = 2 : \mathbb{N}^n \text{ if Cheryl's eyes are green}\\
&\boldsymbol{X} = 3 : \mathbb{N}^n \text{ if Cheryl's eyes are blue}\\
&\quad\ldots\\
&\boldsymbol{X} = n : \mathbb{N}^n \text{ if Cheryl's eyes are}\ldots
\end{aligned}$$

Certainly the question ***Are Cheryl's eyes blue?*** can also be defined over a larger set, if several degrees of colour are to be included as an answer, or alternatively degrees of certainty (definitely blue, quite blue, slightly blue…). Let us assume then another set $\mathbb{N}^j$ for the degree of colour:

$$\begin{aligned}
&\boldsymbol{Y} = 0_1 : \mathbb{N}^j \text{ if Cheryl's eyes are dark blue}\\
&\boldsymbol{Y} = 0_2 : \mathbb{N}^j \text{ if Cheryl's eyes are light blue.}\\
&\boldsymbol{Y} = 0_3 : \mathbb{N}^j \text{ if Cheryl's eyes are green-blue.}\\
&\quad\ldots\\
&\boldsymbol{Y} = 0_j : \mathbb{N}^j \text{ if Cheryl's eyes are}\ldots
\end{aligned}$$

The general dialogical rule for an empirical quantity can thus be rendered:

General dialogical rule for an empirical quantity

	Move	Challenge	Defence
Empirical quantity	**X** X:$\mathbb{N}^n$	**Y** ?= X	**X** m_1 = X:$\mathbb{N}^n$ … **X** m_n = X:$\mathbb{N}^n$ (the defender chooses)

Notice that determining the value of the empirical quantity is an empirical procedure, specific to that quantity; the result of carrying out such a procedure is determined by the rules for that quantity. Moreover, the value of two different empirical quantities might be the same: the quantities only indicate that *the way of determining* the answer to the question might be the same. Take for example these two enquiries

(1) *Did Jorge Luis Borges compose the poem "Ajedrez"?*
(2) *Is Ibn al-Haytham the author of* Al-Shukūk ʿalā Batlamyūs *(Doubts Concerning Ptolemy)?*

These two enquiries involve determining the value of the empirical quantity X for (1) and Y for (2), which can be the same: they can both be ***yes*** for instance if the underlying set is **Bool**.

This leads to a Socratic rule specific to statements of the form **X**, $Y, Z : \mathbb{N}^n$. For example, given the set $\mathbb{N}^n$, **P** can defend the challenges

O$? = X$, with the statement **P** $m_1 = X : \mathbb{N}^n$
O$? = Y$, with the statement **P** $m_2 = Y : \mathbb{N}^n$
O$? = Z$, with the statement **P** $m_3 = Z : \mathbb{N}^n$

Incompatibility

A system of rules that targets the development of a more complex meaning network might include incompatibility rules formulated as challenges. Thus, instead of establishing the simple use of Copy-cat, the game might include more sophisticated rules specific to a particular empirical quantity. For example, if a player responded ***yes*** to the enquiry associated with X

(3) *Did the Greek won in 480 BC the sea-battle take of Salamis*?

that is, if he stated ***yes*** $= X$: **Bool**; this player might not be allowed to respond ***yes*** to the enquiry associated with Z

(4) *Did Xerxes won in 480 BC the sea-battle of Salamis?*

that is, he might not be entitled to further state ***yes*** $= Z$: **Bool**. That is, the other player may challenge the right to answer both (3) and (4) with ***yes***:

(5) *Both answers cannot be* ***yes***.

that is, she can challenge his two statements by stating that $\neg(\mathbf{Id}(\mathbf{Bool}, \mathbf{yes}, X) \wedge \mathbf{Id}(\mathbf{Bool}, \mathbf{yes}, Z))$. The first player would then have to give up. This challenge would be calling upon some *formal incompatibility* between two statements.

Formal incompatibility

	Moves	Challenge	Defence
Formal incompatibility	**P** $yes = X$:**Bool** *and* **P** $yes = Z$:**Bool**	**O** $\neg(Id(\textbf{Bool}, yes, X) \wedge Id(\textbf{Bool}, yes, Z))$	**P** *gives up*

But there is another kind of incompatibility challenge, calling upon *contentual incompatibility*. Consider for instance (4): if a player answers **yes**, *Xerxes won in 480 BC the sea-battle of Salamis*, then the other player can challenge this through contentual incompatibility: the challenger simply states the formally incompatible answer to the challenged statement: $Id(\textbf{Bool}, yes, X)$, *The Greek won in 480 BC the sea-battle of Salamis.* The challenged player must then give up.

Contentual incompatibility

	Moves	Challenge	Defence
Contentual incompatibility	**P** $yes = Z$:**Bool**	**O** $Id(\textbf{Bool}, yes, X)$	**P** *gives up*

Dependent Empirical Quantities

Another more sophisticated form of dealing with empirical quantities is to implement a structure where one empirical quantity might depend on another one. For example let us define the empirical quantity $\boldsymbol{Y}$ as the function $b(\boldsymbol{X}) : \mathbb{N}^n_j[\boldsymbol{X} : \mathbb{N}^n]$ such that

$$\begin{aligned} \boldsymbol{Y} : &=_{df} b(\boldsymbol{X}) : \mathbb{N}^{\mathbf{n}}_j[\boldsymbol{X} : \mathbb{N}^{\mathbf{n}}] \\ & b(\boldsymbol{X}) = j_i : \mathbb{N}^j, \text{ given } \boldsymbol{X} = n_m : \mathbb{N}^{\mathbf{n}} \\ & \ldots \\ & b(\boldsymbol{X}) = j_k : \mathbb{N}^{\mathbf{j}}, \text{ given } \boldsymbol{X} = n_n : \mathbb{N}^{\mathbf{n}}, \text{if} \ldots \end{aligned}$$

Suppose we are interested in determining the meaning of some empirical propositions; this can involve for instance establishing that stating that something has a determinate colour (say, *red*) would presuppose that the player already answered the question whether the object at stake *is coloured or not*.

In this case also the rules of the game might include rules for challenging empirical quantities on the basis of a certain evaluation of another empirical quantity on which the first is dependent; this would be like challenging that something is red by denying that the empirical quantity that yields the evaluation $\boldsymbol{X}$ has a positive response to the question if the object at stake has a colour. We will stop here and invite again the reader to visit the book *Immanent Reasoning*.

The final section of the chapter on *Material dialogues* in Rahman et al. (2018), discusses the epistemological notion of *internalization of content.*[21] In this respect, the dialogical framework can be considered as a formal approach to reasoning rooted in the dialogical constitution and "internalization" of content—including empirical content—rather than in the syntactic manipulation of un-interpreted signs.

This discussion on material dialogues provides a new perspective on Sellars' (1991, pp. 129–194) notion of *Space of Reasons*: the dialogical framework of immanent reasoning enriched with the material level should show how to integrate world-directed thoughts (displaying empirical content) into an inferentialist approach, thereby suggesting that immanent reasoning can integrate within the same epistemological framework the two conflicting readings of the Space of Reasons brought forward by McDowell (2009, pp. 221–238) on the one hand, who insists in distinguishing world-direct thought and knowledge gathered by inference, and Robert Brandom (1997) on the other hand, who interprets Sellars' work in a more radical anti-empiricist manner.

The point is not only that we can deploy the CTT-distinction between *reason* as a premise and *reason* as a piece of evidence justifying a proposition, but also that the dialogical framework allows for distinguishing between the **objective justification level** targeted by Brandom (1997, p. 129) and the **subjective justification level** stressed by McDowell.

According to our approach the subjective feature corresponds to the play level, where a concrete player brings forward the statement *It looks red to me*, rather than *It is red*. The general epistemological upshot from these initial reflections is that, on our view, many of the worries on the interpretation of the *Space of Reasons* and on the shortcomings of the standard dialogical approach to meaning have their origin in the neglect of the play-level.

2.3 Strategic Reasons in Dialogues for Immanent Reasoning

The conceptual backbone on which rests the metalogical properties of the dialogical framework is the notion of strategic reason which allows to adopt a global view on all the possible plays that constitute a strategy. However, this global view should not be identified with the perspective common in proof theory: strategic reasons are a kind of recapitulation of what can happen for a given thesis and show the entire history of the play by means of the instructions. Strategic reasons thus yield an overview of the possibilities enclosed in a thesis—what plays can be carried out from it—, but without ever being carried out in an actual play: they are only a perspective on all the possible variants of plays for a thesis and not an actual play. In this way the rules

[21] By "internalization" we mean that the relevant content is made part of the setting of the game of giving and asking for reasons: any relevant content is the content displayed during the interaction. For a discussion on this conception of internalization—see Peregrin (2014, pp. 36–42).

of synthesis and analysis of strategic reasons provided below are not of the same nature as the analysis and synthesis of local reasons, they are not produced through challenges and their defence, but are a recapitulation of the plays that can actually be carried out.

The notion of strategic reasons enables us to link dialogical strategies with CTT-demonstrations, since strategic reasons (and not local reasons) are the dialogical counterpart of CTT proof-objects; but it also shows clearly that the strategy level by itself—the only level that proof theory considers—is not enough: a deeper insight is gained when considering, together with the strategy level, the fundamental level of plays; strategic reasons thus bridge these two perspectives, the global view of strategies and the more in-depth and down-to-earth view of actual plays with all the possible variations in logic they allow,[22] without sacrificing the one for the other.

This vindication of the play level is a key aspect of the dialogical framework and one of the purposes of the present study: other logical frameworks lack this dimension, which besides is not an extra dimension appended to the concern for demonstrations, but actually constitutes it, the heuristic procedure for building strategies out of plays showing the gapless link there is between the play level and the strategy level: strategies (and so demonstrations) stem from plays. Thus the dialogical framework can say at least as much as other logical frameworks, and, additionally, reveals limitations of other frameworks through this level of plays.

2.3.1 Introducing Strategic Reasons

Strategic reasons belong to the strategy level, but are elements of the object-language of the play level: they are the reasons brought forward by a player entitling him to his statement. Strategic reasons are a perspective on plays that take into account all the possible variations in the play for a given thesis; they are never actually carried out, since any play is but the actualization of only one of all the possible plays for the thesis: each individual play can be actualized but will be separate from the other individual plays that can be carried out if other choices are made; strategic reasons allow to see together all these possible plays that in fact are always separate. There will never be in any of the plays the complex strategic reason for the thesis as a result of the application of the particle rules, only the local reason for each of the subformulas involved; the strategic reason will put all these separate reasons together as a recapitulation of what can be said from the given thesis.

Consider for instance a conjunction: the Proponent claims to have a strategic reason for this conjunction. This means that he claims that whatever the Opponent might play, be it a challenge of the left or of the right conjunct, the Proponent will be able to win the play. But in a single play with repetition rank 1 for the

[22] Among these variations can be counted cooperative games, non-monotony, the possibility of player errors or of limited knowledge or resources, to cite but a few options the play level offers, making the dialogical framework very well adapted for history and philosophy of logic.

Opponent, there is no way to check if a conjunction is justified, that is if both of the conjuncts can be defended, since a play is precisely the carrying out of only one of the possible **O**-choices (challenging the left or the right conjunct): to check both sides of a conjunction, two plays are required, one in which the Opponent challenges the left side of the conjunction and another one for the right side. So a strategic reason is never a single play, but refers to the strategy level where all the possible outcomes are taken into account; the winning strategy can then be displayed as a tree showing that both plays (respectively challenging and defending the left conjunct and right conjunct) are won by the Proponent, thus justifying the conjunction.

Let us now study what strategic reasons look like, how they are generated and how they are analyzed.

A strategic *perspective* on a statement

In the standard framework of dialogues, where we do not explicitly have the reasons for the statements in the object-language, the particle rules simply determine the local meaning of the expressions. In dialogues for immanent reasoning, the reasons entitling one to a statement are explicitely introduced; the particle rules (synthesis and analysis of local reasons) govern both the local reasons and the local meaning of expressions. But when building the core of a winning **P**-strategy, local reasons are also linked to the justification of the statements—which is not the case if considering single plays, for then only one aspect of the statement may be taken into account during the play, the play providing thus only a partial justification.

Take again the example of a **P**-conjunction, say

$$\mathbf{P}w : A \wedge B.$$

In providing a strategic reason w for the conjunction $A \wedge B$, **P** is claiming to have a winning strategy for this conjunction, that is, he is claiming that the conjunction is absolutely justified, that he has a proper reason for asserting it and not simply a *local* reason for stating it. Assuming that **O** has a repetition rank of 1 and has stated both A and B prior to move i, two different plays can be carried out from this point, which we provide without the strategic reason:

Introducing strategic reasons: stating a conjunction

O			**P**		
	Concessions			Thesis	0
1	$m := 1$			$n := 2$	2
...	...			...	...
...	...			$!A \wedge B$	i

Introducing strategic reasons: left decision option on conjunction

O			**P**		
	Concessions			Thesis	0
1	$m := 1$			$n := 2$	2
…	…			…	…
…	…			$!A \wedge B$	i
$i+1$	$?\wedge_1$	i		$!A$	$i+2$

Introducing strategic reasons: right decision option on conjunction

O			**P**		
	Concessions			Thesis	0
1	$m := 1$			$n := 2$	2
…	…			…	…
…	…			$!A \wedge B$	i
$i+1$	$?\wedge_2$	i		$!B$	$i+2$

So if **P** brings forward the strategic reason *w* to support his conjunction at move *i*, he is claiming to be able to win *both* plays, and yet the actual play will follow into only one of the two plays. Strategic reasons are thus a *strategic perspective* on a statement that is brought forward during actual plays.

An anticipation of the play and strategy as recapitulation

Since a strategic reason (*w* for instance) is brought forward during a play (say at move *i*), it is clear that the play has not yet been carried out fully when the player claims to be able to defend his statement against whatever challenge his opponent might launch: bringing forward a strategic reason is thus an anticipation on the outcome of the play.

But strategic reasons are not a simple claim to have a winning strategy, they also have a complex internal structure: they can thus be considered as recapitulations of the plays of the winning strategy produced by the heuristic procedure, that is the winning strategy obtained only after running all the relevant plays; this strategy-building process specific to the dialogical framework is a richer process than the one yielding CTT demonstrations—or proof theory in general—, since the strategic reasons will contain traces of *choice dependences*, which constitute their complexity.

Choice dependences link possible moves of a player to the choices made by the other player: a player will play this move if his opponent used this decision-option, that move if the opponent used that decision-option. In the previous example, the Proponent will play move $i + 2$ depending on the Opponent's decision at move $i + 1$, so the strategic object w played at move i will contain these two possible scenarios with the $i + 2$ **P**-move depending of the $i + 1$ **O**-decision. The strategic reason w is thus a *recapitulation* of what would happen if each relevant play was carried out. When the strategic reason makes clearly explicit this choice-dependence of **P**'s moves on those of **O**, we say that it is in a *canonical argumentation form* and is a recapitulation of the statement.

The rules for strategic reasons do not provide the rules on how to play but rather rules that indicate how a winning strategy has been achieved while applying the relevant rules at the play level. Strategic reasons emerge as the result of considering the optimal moves for a winning strategy: this is what a recapitulation is about.

The canonical argumentation form of strategic reasons is closely linked to the synthesis and analysis of local reasons: they provide the recapitulation of all the relevant local reasons that could be generated from a statement. In this respect following the rules for the synthesis and analysis of local reasons, the rules for strategic reasons are divided into synthesis and analysis of strategic reasons, to which we will now turn.

In a nutshell, the synthesis of strategic reasons provides a guide for what **P** needs to be able to defend in order to justify his claim; the analysis of strategic reasons provides a guide for the local reasons **P** needs to make **O** state in order to copy these reasons and thus defend his statement.

Assertions and statements

The difference between local reasons and strategic reasons should now be clear: while local reasons provide a local justification entitling one to his statement, strategic reasons provide an absolute justification of the statement, which thus becomes an *assertion*.

The equalities provided in each of the plays constituting a **P**-winning strategy, and found in the analysis of strategic objects, convey the information required for **P** to play in the best possible way by specifying those **O**-moves necessary for **P**'s victory. This information however is not available at the very beginning of the first play, it is not made explicit at the root of the tree containing all the plays relevant for the **P**-winning strategy: the root of the tree will not explicitly display the information gathered while developing the plays; this information will be available only once the whole strategy has been developed, and each possible play considered. So when a play starts, the thesis is a simple statement; it is only at the end of the construction process of the strategic reason that **P** will be able to have the knowledge required to assert the thesis, and thus provide in any new play a strategic reason for backing his thesis.

The *assertion* of the thesis, making explicit the strategic reason resulting from the plays, is in this respect a *recapitulation* of the result achieved after running the relevant plays, after **P**'s initial simple statement of that thesis. This is what the canonical argumentation form of a strategic object is, and what renders the dialogical formulation of a CTT canonical proof-object.

It is in this fashion that dialogical reasons correspond to CTT proof-objects: introduction rules are usually characterized as the right to assert the conclusion from the premises of the inference, that is, as defining what one needs in order to be entitled to assert the conclusion; and the elimination rules are what can be inferred from a given statement. Thus, in the dialogical perspective of **P**-winning strategies, since we are looking at **P**'s entitlements and duties, what corresponds to proof-object introduction rules would define what **P** is required to justify in order to assert his statement, which is the synthesis of a **P**-strategic reason; and what corresponds to proof-object elimination rules would define what **P** is entitled to ask of **O** from her previous statements and thus say it himself by copying her statements, which is the analysis of **P**-strategic reasons. We will thus provide the rules for the synthesis and analysis of strategic reasons (always in the perspective of a **P**-winning strategy), followed by their corresponding CTT rule. We have in this regard a good justification of Sundholm's idea that inferences can be considered as involving an *(implicit) interlocutor*, but here at the strategy level.

2.3.2 *Rules for the Synthesis of P-Strategic Reasons*

P-strategic reasons must be built (*synthesis* of **P**-strategic reasons); they constitute the justification of a statement by providing certain information—choice-dependences—that is essential to the relevant plays issuing from the statement: strategic reasons are a recapitulation of the building of a winning strategy, directly inserted into a play. Thus a strategic reason for a **P**-statement on the universal **P** ! $(\forall x\colon A)\,B(x)$ has the form $\lambda(x^{\mathbf{O}})b^{\mathbf{P}}(x)$, which indicates that **P** has some method $b(x)$, that delivers a winning strategy for $B(x)$,whatever local reason x **O** choses for stating the antecedent. Moreover, it indicates that **P**'s choice for defending the right constituent (its consequent) of the universal is dependent upon **O**'s choice for stating the antecedent.

Strategic reasons for **P** are the dialogical formulation of CTT proof-objects, and the canonical argumentation form of strategic reasons correspond to canonical proof-objects.

Synthesis of strategic reasons for P:

	Move	Synthesis of local reasons		Synthesis of strategic reasons Canonical Argumentation form
		Challenge	Defence	
Conjunction	**P** ! $A \wedge B$	**O** ?$L^{\wedge}$ or **O** ?$R^{\wedge}$	**P** p_1:A (resp.) **P** p_2:B	**P** $< p_1, p_2 >$:$A \wedge B$
Existential quantification	**P**! $(\exists x{:}A)B(x)$	**O** ?$L^{\exists}$ or **O** ?$R^{\exists}$	**P** p_1:A (resp.) **P** p_2:$B(p_1)$	**P** $< p_1, p_2 >$:$(\exists x{:}A)B(x)$
Disjunction	**P** !$A \vee B$	**O** ?$^{\vee}$	**P** p_1:A or **P** p_2:B	**P** **i**(p_1):$A \vee B$ or **P** **j**(p_2):$A \vee B$ The strategic reason **i**(p_1) indicates that **P** has chosen, the left side to build a winning-strategy for the disjunction – **i**(p_1) amounts to **P**. choosing p_1 as strategic reason for the disjunction. Analogous holds for **j**(p_2), that indicates **P**'s choice for the right-side
Implication	**P** ! $A \supset B$	**O** p_1:A	**P** p_2:B	**P** $\lambda(x^{\mathbf{O}})b^{\mathbf{P}}(x)$:$A \supset B$ $\lambda(x^{\mathbf{O}})b^{\mathbf{P}}(x)$ indicates that **P** has some method $b(x)$, which delivers a winning strategy for the consequent whatever local reason x **O** choses for stating the antecedent
Universal quantification	**P** ! $(\forall x{:}A)B(x)$	**O** p_1:A	**P** p_2:$B(p_1)$	**P** $\lambda(x^{\mathbf{O}})b^{\mathbf{P}}(x)$:$(\forall x{:}A)B(x)$ $\lambda(x^{\mathbf{O}})b^{\mathbf{P}}(x)$ indicates that **P** has some method $b(x)$, which delivers a winning strategy for $B(x)$, whatever local reason x **O** choses for stating the antecedent
Negation	**P** ! $A \supset \bot$	**O** p_1:A … **O** ! $\bot$ (stating the antecedent leads eventually to **O** giving up)	——	**P** $\lambda(x^{\mathbf{O}})b^{\mathbf{P}}(x)$:$A \supset \bot$ The method $b(x)$ encoded by $\lambda(x^{\mathbf{O}})b^{\mathbf{P}}(x)$ will never be carried out. Indeed, since $\lambda(x^{\mathbf{O}})b^{\mathbf{P}}(x)$ provides a winning strategy, **P** will force **O** to state **falsum** himself (on the grounds of the move **O** p_1:A), before $b(x)$ comes into play

Remark: For the case of negation, we must bear in mind that we are considering **P**-strategies, that is, plays in which **P** wins, and we are not providing particle rules with a proper challenge and defence, but we are adopting a strategic perspective on the reason to provide backing a statement; thus the response to an **O**-challenge on a negation cannot be **P** ! $\bot$, which would amount to **P** losing

2.3.3 Rules for the Analysis of P-Strategic Reasons

Analysis rules for P-strategic reasons

	Move	Analysis of local reasons		Analysis of P-strategic reasons
		Challenge	Defence	
Conjunction	**O** $p{:}A \wedge B$	**P** ? $L^{\wedge}$ or **P** ? $R^{\wedge}$	**O** $L^{\wedge}(p){:}A$ … **O** $L^{\wedge}(p) = p_1{}^{\mathbf{O}}{:}A$ (resp.) **O** $R^{\wedge}(p){:}B$ … **O** $R^{\wedge}(p) = p_2{}^{\mathbf{O}}{:}B$	**O** $p{:}A \wedge B$ **P** ! $C(p)$) ⇓ **P** $c{:}C(\mathbf{fst}(p))$ (resp.) **O** $p{:}A \wedge B$ **P** ! $C(p)$) ⇓ **P** $d{:}C(\mathbf{snd}(p))$ The rules indicate that, given **O** $p{:}A \wedge B$, **P**'s winning strategy for $C(p)$ is dependent upon **O**'s choice for a local reason for the first (or second) element of the conjunction Where **O** $\mathbf{fst}(p)= p_1{}^{\mathbf{O}}{:}A$, and Where **O** $\mathbf{snd}(p)= p_2{}^{\mathbf{O}}{:}B$
Existential quantification	**O** $p{:}(\exists x{:}A)B(x)$	**P** ? $L^{\exists}$ or **P** ? $R^{\exists}$	**O** $L^{\exists}(p){:}A$ … **O** $L^{\exists}(p) = p_1{}^{\mathbf{O}}{:}A$ (resp.) **O** $R^{\exists}(p){:}B(L^{\exists}(p)^{\mathbf{O}})$ … **O** $R^{\exists}(p) = p_2{}^{\mathbf{O}}{:}B(p_1{}^{\mathbf{O}})$	**O** $p{:}(\exists x{:}A)B(x)$ **P** ! $C(p)$) ⇓ **P** $c{:}C(\mathbf{fst}(p))$ (resp.) **O** $p{:}(\exists x{:}A)B(x)$ **P** ! $C(p)$) ⇓ **P** $d{:}C(\mathbf{snd}(p))$ Where **O** $\mathbf{fst}(p)= p_1{}^{\mathbf{O}}{:}A$, and Where **O** $\mathbf{snd}(p)= p_2{}^{\mathbf{O}}{:}B(p_1{}^{\mathbf{O}})$
Disjunction	**O** $p{:}A \vee B$	**P** $?^{\vee}$	**O**$L^{\vee}(p\ ){:}A$ … **O** $L^{\vee}(p) = p_1{}^{\mathbf{O}}{:}A$ or **O** $R^{\vee}(p\ ){:}B$ … **O** $R^{\vee}(p) = p_2{}^{\mathbf{O}}{:}B$	**O** $p{:}\ A \vee B$ **O** $x^{\mathbf{O}}{:}A$ / **O** $y^{\mathbf{O}}{:}B$ **P** ! $C(\mathbf{i}(x^{\mathbf{O}})$ / **P** ! $C(\mathbf{j}(y^{\mathbf{O}})$ ⇓ / ⇓ **P** $d{:}C(\mathbf{i}(x^{\mathbf{O}})$ / **P** $e{:}C(\mathbf{j}(y^{\mathbf{O}})$ ⇓ **P** $\mathbf{D}(p, x^{\mathbf{O}}.d; y^{\mathbf{O}}.e){:}C(p)$ The strategic reason $\mathbf{D}(p, x^{\mathbf{O}}.d; y^{\mathbf{O}}.e)$ indicates that **P**'s winning strategy for $C(p)$, is dependent upon **O**'s choices for the disjunction. In other words **P** $\mathbf{D}(\mathbf{i}(p_1), x^{\mathbf{O}}.d; y^{\mathbf{O}}.e) = d[p_1]{:}C(\mathbf{i}(p_1))$. Similarly **P** $\mathbf{D}(\mathbf{j}(p_2), x^{\mathbf{O}}.d; y^{\mathbf{O}}.e) = e[p_2]{:}C(\mathbf{j}(p_2))$

Implication	**O** $p{:}A \supset B$	**P** $L^{\supset}(p){:}A$ ... **P** $L^{\supset}(p) = p_1{}^{\mathbf{P}}$:	**O** $R^{\supset}(p){:}B$ ... **O** $R^{\supset}(p) = p_2{:}B$	**O** $p{:}A \supset B$ **O** $p_1{:}A$ **P** ! B ⇓ **P** $p_1{:}A$ ⇓ **P** **ap**$(p, p_1{}^{\mathbf{P}}){:}B$ The strategic reason **ap**$(p, p_1{}^{\mathbf{P}})$ indicates that **P** can build a winning strategy for his claim **P** ! B, given that **O** stated p_1: A and that on that ground **P** can deploy this assertion for a challenge on the implication, in order to force **O** to state the consequent. In other words: **P** **ap**$(p, p_1{}^{\mathbf{P}}) = p_2{}^{\mathbf{O}}{:}B$
Universal quantification	**O** $p{:}(\forall x{:}A)B(x)$	**P** $L^{\forall}(p){:}A$ ... **P** $L^{\forall}(p) = p_1{}^{\mathbf{P}}{:}A$	**O** $R^{\forall}(p){:}B(L^{\forall}(p)^{\mathbf{P}})$ ... **O** $R^{\forall}(p) = p_2{:}B(p_1{}^{\mathbf{P}})$	**O** $p{:}(\forall x{:}A)B(x)$ **O** $p_1{:}A$ **P** ! $B(p_1)$ ⇓ **P** $p_1{:}A$ ⇓ **P** **ap**$(p, p_1{}^{\mathbf{P}}){:}B(p_1)$ The strategic reason **ap**$(p, p_1{}^{\mathbf{P}})$ indicates that **P** can build a winning strategy for his claim **P** ! $B(p_1)$, given that **O** stated p_1: A and that on that ground **P** can deploy this assertion for a challenge on the universal, in order to force **O** to state the consequent. In other words: **P** **ap**$(p, p_1{}^{\mathbf{P}}) = p_2{}^{\mathbf{O}}{:}B(p_1)$
Negation	**O** $p{:}A \supset \bot$	**P** $L^{\supset}(p){:}A$ ... **P** $L^{\supset}(p) = p_1{}^{\mathbf{P}}$:	**O** $R^{\supset}(p){:}\bot$ ... The instruction $R^{\supset}(p)$ keeps un-resolved	**O** $p{:}A \supset \bot$ **O** $p_1{:}A$ **P** ! C ⇓ **P** $p_1{:}A$ ⇓ **O** **ap**$(p, p_1{}^{\mathbf{P}}){:}\bot$ ⇓ **P** $you_{gave\ up}(\mathrm{n}){:}C$ The strategic reason **ap**$(p, p_1{}^{\mathbf{P}})$ indicates that **P** can build a winning strategy for C, given that **O** stated p_1: A and that on that ground, **P** can deploy this assertion for a challenge on the implication, in order to force **O** to give up by stating **falsum**

2.4 A Plaidoyer for the Play-Level

To some extent, the criticisms the dialogical approach to logic has been subject to provides an opportunity for clarifying its basic tenets. We will therefore herewith consider some recent objections raised against the dialogical framework in order to pinpoint some of its fundamental features, whose importance may not have appeared clearly enough through the main body of the paper; namely,

dialogue-definiteness,
player-independence, and

the dialogical conception of proposition.

Showing how and why these features have been developed, and specifying their point and the level they operate on, will enable us to vindicate the play level and thus disarm the objections that have been raised against the dialogical framework for having neglected this crucial level.

We shall first come back on the central notion of dialogue-definiteness and on the dialogical conception of propositions, which are essential for properly understanding the specific role and importance of the play level. We shall then be able to address three objections to the dialogical framework, due to a misunderstanding of the notion of *Built-in Opponent*, of the principles of dialogue-definiteness and of player-independence, and of the reflection on normativity that constitutes the philosophical foundation of the framework; all of these misunderstandings can be reduced to a misappraisal of the play level. We shall then go somewhat deeper in the normative aspects of the dialogical framework, according to the principle that logic has its roots in ethics.

2.4.1 Dialogue-Definiteness and Propositions

The dialogical theory of meaning is structured in three levels, that of the local meaning (determined by the particle rules for the logical constants), of the global meaning (determined by the structural rules), and the strategic level of meaning (determined by what is required for having a winning strategy). The material level of consideration is part of the global meaning, but with particular rules so precise that they determine only one specific expression (through a modified Socratic rule). A characteristic of the local meaning is that the rules are player independent: the meaning is thus defined in the same fashion for each player; they are bound by the same sets of duties and rights when they start a dialogue. This normative aspect is thus constitutive of the play level (which encompasses both the local meaning and the global meaning): it is even what allows one to judge that a dialogue is taking place. In this regard, meaning is immanent to the dialogue: what constitutes the meaning of the statements in a particular dialogue solely rests on rules determining interaction (the local and the global levels of meaning). The strategy level on the other hand is built on the play level, and the notion of demonstration operates on the strategy level (it amounts to having a winning strategy).

Two main tenets of the dialogical theory of meaning can be traced back to Wittgenstein, and ground in particular the pivotal notion of dialogue-definiteness:

1. the internal feature of meaning (the *Unhintergehbarkeit der Sprache*[23]), and
2. the meaning as mediated by language-games.

As for the first Wittgensteinian tenet, the internal feature of meaning, we already mentioned in the introduction that if we relate the notion of internalization of meaning

[23]See *Tractatus Logico-Philosophicus,* 5.6.

with both language-games and fully-interpreted languages of CTT, then a salient feature of the dialogical approach to meaning can come to fore: the expressive power of CTT allows all these actions involved in the dialogical constitution of meaning to be incorporated as an explicit part of the object-language of the dialogical framework.

In relation to the second tenet, the inceptors of the dialogical framework observed that if language-games are to be conceived as mediators of meaning carried out by social interaction, these language-games must be games actually playable by human beings: it must be the case that *we can actually perform them*,[24] which is captured in the notion of dialogue-definiteness.[25] Dialogue-definiteness is essential for dialogues to be mediators of meaning, but it is also constitutive of what propositions are, as Lorenz clearly puts it:

> [...] for an entity to be a proposition there must exist an individual play, such that this entity occupies the initial position, and the play reaches a final position with either win or loss after a finite number of moves according to definite rules (Lorenz 2001, p. 258).

A proposition is thus defined in the standard presentation of dialogical logic as a dialogue-definite expression, that is, an expression *A* such that there is an individual play about *A*, that can be said to be lost or won after a finite number of steps, following given rules of dialogical interaction.[26]

The notion of *dialogue-definiteness* is in this sense the backbone of the dialogical theory of meaning: it provides the basis for implementing the human-playability requirement and the notion of proposition.

Dialogue-definiteness sets apart rather decisively the level of strategies from the level of plays, as Lorenz's notion of dialogue-definite proposition does not amount to a set of winning strategies, but rather to an individual play. Indeed, a winning strategy for a player **X** is a sequences of moves such that **X** wins *independently of the moves of the antagonist*. It is crucial to understand that the qualification *independently of the moves of the antagonist* amounts to the fact that the one claiming *A* has to play under the restriction of the Copy-cat rule: if possessing a winning strategy

---

[24]As observed by Marion (2006, p. 245), a lucid formulation of this point is the following remark of Hintikka (1996, p. 158) who shared this tenet (among others) with the dialogical framework:

> [Finitism] was for Wittgenstein merely one way of defending the need of language-games as the sense that [sic] they had to be actually playable by human beings. [...] Wittgenstein shunned infinity because it presupposed constructions that we human beings cannot actually carry out and which therefore cannot be incorporated in any realistic language-game. [...] What was important for Wittgenstein was not just the finitude of the operations we perform in our calculi and other language-games, but the fact that we can actually perform them. Otherwise the entire idea of language-games as meaning mediators will lose its meaning. The language-games have to be humanly playable. And that is not possible if they involve infinitary elements. Thus it is the possibility of actually playing the meaning-conferring language-games that is the crucial issue for Wittgenstein, not finitism as such.

[25]The fact that these language-games must be finite does not rule out the possibility of a (potentially) infinite number of them.

[26]While establishing particle rules the development rules have not been fixed yet, so we might call those expressions *propositional schemata*.

for player **X** involves being in possession of a method (leading to the win of **X**) allowing to choose a move for any move the antagonist might play, then we must assume that the propositions brought forward by the antagonist are justified. There is a winning strategy if **X** can base his moves leading to a win by endorsing himself those propositions whose justification is rooted on **Y**'s authority. For short, the act of endorsing is what lies behind the so-called Copy-cat rule and structures dialogues for immanent reasoning: it ensures that **X** can win whatever the contender might bring forward in order to contest A (within the limits set by the game).

Furthermore, *refuting*, that is bringing up a strategy *against* A, amounts to the dual requirement: that the antagonist **Y** possess a method that leads to the loss of **X** $!A$, whatever **X** is can bring forward, and that she can do it under the Copy-cat restriction:

> **X** $!A$ is refuted, if the antagonist **Y** can bring up a sequence of moves such that she (**Y**) can win playing under the Copy-cat restriction.

Refuting is thus different and stronger than contesting: while *contesting* only requires that the antagonist **Y** brings forward at least one counterexample in a kind of play where **Y** does not need to justify her own propositions, *refuting* means that **Y** must be able to lead to the loss of **X** $!A$, *whatever X's justification of his propositions might be*.

In this sense, the assumption that every play is a finitary open two-person zero-sum game does not mean that either there is a winning strategy for A or a winning strategy *against* A : the play level cannot be reduced to the strategy level.

For instance, if we play with the Last-duty first development rule **P** will lose the individual plays relevant for the constitution of a strategy for $\vee\neg A$. So $A \vee \neg A$ is *dialogue-definite*, though there is no winning strategy *against* $A \vee \neg A$.

The distinction between the play level and the strategy level thus emerges from the combination of dialogue-definiteness and the Copy-cat rule.

The classical reduction of strategies against A to the falsity of A (by means of the saddle-point theorem) assumes that the win and the loss of a *play* reduce to the truth or the falsity of the thesis. But we claim that the existence of the play level and a loss in one of the plays introduces a qualification that is not usually present in the purely proof-theoretic approach; to use the previous example, we know that **P** does not have a winning strategy for $!A \vee \neg A$ (playing under the intuitionisitic development rule), but neither will **O** have one against it if she has to play under the Copy-cat rule herself (notice the switch in the burden of the restriction of the Copy-cat rule when *refuting* a thesis). Let us identify the player who has to play under the Copy-cat restriction by highlighting her moves:

Play *against* **P** ! $A \vee \neg A$

O			**P**		
				! $A \vee \neg A$	0
1	$n := 1$			$m := 2$	2
3	$?_{\vee}$	0		! A	4
				P wins	

The distinction between the play and the strategy level can be understood as a consequence of introducing the notion of dialogue-definiteness which amounts to a win or a loss at the play level, though strategically seen, the proposition at stake may be (proof-theoretically) undecidable. Hence, some criticisms to the purported lack of dynamics to dialogical logic are off the mark if they are based on the point that "games" of dialogical logic are deterministic[27]: plays are deterministic in the sense that they are dialogue-definite, but strategies are not deterministic in the sense that for every proposition there would either be a winning strategy for it or a winning strategy *against* it.

Before ending this section let us quote quite extensively (Lorenz 2001), who provides a synopsis of the historical background that lead to the introduction of the notion of dialogue-definiteness and the distinction of the deterministic conception of plays—which obviously operates at the level of plays—from the proof-theoretical undecidable propositions—which operate at the level of strategies:

> [...] It was Alfred Tarski who, in discussions with Lorenzen in 1957/58, when Lorenzen had been invited to the Institute for Advanced Study at Princeton, convinced him of the impossibility to characterize arbitrary (logically compound) propositions by some decidable generalization of having a decidable proof-predicate or a decidable refutation-predicate.
>
> [...] It became necessary to search for some decidable predicate which may be used to qualify a linguistic entity as a proposition about any domain of objects, be it elementary or logically compound. Decidability is essential here, because the classical characterization of a proposition as an entity which may be true or false, has the awkward consequence that of an undecided proposition it is impossible to know that it is in fact a proposition. This observation gains further weight by L. E. J. Brouwer's discovery that even on the basis of a set of "value-definite", i.e., decidably true or false, elementary propositions, logical composition does not in general preserve value-definiteness. And since neither the property of being proof-definite nor the one of being refutation-definite nor properties which may be defined using these two, are general enough to cover the case of an arbitrary proposition, some other procedure had to be invented which is both characteristic of a proposition and satisfies a decidable concept. The concept looked for and at first erroneously held to be synonymous with argumentation [[28]] turned out to be the concept of dialogue about a proposition A (which had to replace

---

[27]For such criticisms—see Trafford (2017, pp. 86–88).

[28]Lorenz identifies argumentation rules with rules at the strategy level and he would like to isolate the interaction displayed by the moves constituting the play level — see Lorenz (2010, p.79). We deploy the term argumentation-rule for request-answer interaction as defined by the local and structural rules. It is true that nowadays argumentation-rules has even a broader scope including several kinds of communicative interaction and this might produce some confusion on the main goal of the dialogical framework which is in principle, to provide an argumentative understanding of logic rather than the logic of argumentation. However, once this distinction has been drawn nothing

> the concept of truth of a proposition A as well as the concepts of proof or of refutation of a proposition A, because neither of them can be made decidable). Fully spelled out it means that for an entity to be a proposition there must exist a dialogue game associated with this entity, i.e., the proposition A, such that an individual play of the game where A occupies the initial position, i.e., a dialogue D(A) about A, reaches a final position with either win or loss after a finite number of moves according to definite rules: the dialogue game is defined as a finitary open two-person zero-sum game. Thus, propositions will in general be dialogue-definite, and only in special cases be either proof-definite or refutation-definite or even both which implies their being value-definite. Within this game-theoretic framework where win or loss of a dialogue D(A) about A is in general not a function of A alone, but is dependent on the moves of the particular play D(A), truth of A is defined as existence of a winning strategy for A in a dialogue game about A; falsehood of A respectively as existence of a winning strategy against A. Winning strategies for A count as proofs of A, and winning strategies against A as refutations of A. The meta-truth of "either 'A is true' or 'A is false' " which is provable only classically by means of the saddlepoint theorem for games of this kind may constructively be reduced to the decidability of win or loss for individual plays about A. The concept of truth of dialogue-definite propositions remains finitary, and it will, as it is to be expected of any adequate definition of truth, in general not be recursively enumerable. The same holds for the concept of falsehood which is conspicuously defined independently of negation (Lorenz 2001, pp. 257–258).

2.4.2 The Built-in Opponent and the Neglect of the Play Level

In recent literature Duthil-Novaes (2015) and Trafford (2017, pp. 102–105) deploy the term *internalization* for the proposal that natural deduction can be seen as having an internalized Opponent, thereby motivating the inferential steps. This form of internalization is called the *built-in Opponent*. The origin of this concept is linked to Göran Sundholm who, by 2000, in order to characterize the fundamental links between natural deduction and dialogical logic, suggested in his lectures and talks the idea that elimination rules can be read as the moves of an Opponent aimed at testing the thesis. Yet, since this reading was meant to link the strategy level with natural deduction, the concept of *built-in Opponent* inherited the same strategic perspective on *logical truth*. Thus, logical truth can be seen as the encoding of a process through which the Proponent succeeds in defending his assertion against a stubborn *ideal* interlocutor.[29]

From the dialogical point of view however, the ideal interlocutor of the strategy level is the result of a process of selecting the relevant moves from the play level. Rahman et al. (2009), in a paper dedicated to the Festschrift for Sundholm, designate the process as *incarnation*, using Jean-Yves Girard's term. Their thorough description of the incarnation process already displays those aspects of the *cooperative*

prevents to develop the interface dialogical-understanding of logic/logical structure of a dialogue. In fact, it is our claim that in order to study the logical structure of a dialogue, the dialogical conception of logic provides the right venue.

[29]With "ideal" we mean an interlocutor that always make the optimal choices in order to collaborate in the task of testing the thesis.

endeavour, which was formulated by Duthil Novaes (2015) and quoted by Trafford (2017, p. 102) as a criticism of the dialogical framework. Their criticism seems to rest on the idea that the dialogues of the dialogical framework are not truly cooperative, since they are reduced to constituting logical truth. If this is really the point of their criticism, it is simply wrong, for the play level would then be completely neglected: the intersubjective in-built and implicit cooperation of the *strategy* level (which takes care of inferences) grows out of the *explicit* interaction of players at the *play* level in relation to the formation-rules; accepting or contesting a local reason is a process by the means of which players cooperate in order to determine the meaning associated to the action-schema at stake.[30]

It is fair to say that the standard dialogical framework, not enriched with the language of CTT, did not have the means to fully develop the so-called material dialogues, that is dialogues that deal with content. Duthil Novaes (2015, p. 602)—but not Trafford (2017, p. 102)—seems to be aware that dialogues are a complex interplay of adversarial and cooperative moves,[31] even in Lorenzen and Lorenz' standard formulation. However, since she understands this interplay as triggered by the built-in implicit Opponent at the strategy level, Duthil Novaes suggestions or corrections motivated by reflections on the Opponent's role cannot be made explicit in the framework.[32]

Duthil Novaes' (2015, pp. 602–604) approach leads her to suggest that monotonicity is a consequence of the role of the Opponent as a stubborn adversary, which

---

[30]In fact, when Trafford (2017) criticizes dialogical logic in his Chap. 4, he surprisingly claims that this form of dialogical interaction does not include the case in which the plays would be open-ended in relation to the logical rules at stake, though it has already been suggested—see for instance in (Rahman and Keiff 2005, pp. 394–403)—how to develop what we called *Structure Seeking Dialogues* (SSD). Moreover, Keiff's (2007) Ph.D.-dissertation is mainly about SSD. The idea behind SSD is roughly the following; let us take some inferential practice we would like to formulate as an action-schema, mainly in a teaching-learning situation; we then search for the rules allowing us to make these inferential practices to be put into a schema. For example: we take the third excluded to be in a given context a sound inferential practice; we then might ask what kind of moves **P** should be allowed to make if he states the third excluded as thesis. It is nonetheless true to say that SSD were studied only in the case of modal logic.

[31]To put it in her own words: "*the majority of dialogical interactions involving humans appear to be essentially cooperative, i.e., the different speakers share common goals, including mutual understanding and possibly a given practical outcome to be achieved.*" Duthil Novaes (2015, p. 602).

[32]See for instance her discussion of countermoves Duthil Novaes (2015, p. 602) : indefeasibility means that the Opponent has no available countermove: "A countermove in this case is the presentation of one single situation, no matter how far-fetched it is, where the premises are the case and the conclusion is not—a counterexample." The question then would be to know how to show that the Opponent has no countermove available. The whole point of building winning strategies from plays is to *actually construct* the evidence that there is no possible move for the Opponent that will lead her to win: that is a winning strategy. But when the play level is neglected, the question remains: how does one know the Opponent has no countermove available? It can actually be argued that the mere notion of countermove tends to blur the distinction between the level of plays and of strategies: a *counter*move makes sense if it is 'counter' to a winning strategy, as if the players were playing at the strategy level, but that is something we explicitly reject. At the play level, there are only simple *moves*: these can be challenges, defences, counterattacks, but *countermoves* do not make any sense.

takes care of the non-defeasibility of the demonstration at stake; from this perspective, she contends that the standard presentations of dialogical logic, being mostly adversarial or competitive, are blind to defeasible forms of reasons and are thus

> [...] rather contrived forms of dialogical interaction, and essentially restricted to specific circles of specialists (Duthil Novaes 2015, p. 602).

But this argument is not compelling when considering the strategy level as being built from the play level: setting aside the point on content mentioned above, if we conceive the constitution of a strategy as the end-result of the complementary role of competition and cooperation taking place at the play level, we do not seem to need—at least in many cases—to endow the notion of inference with non-monotonic features. The play level is the level were cooperative interaction, either constructive or destructive, can take place until the definitive answer—given the structural and material conditions of the rules of the game—has been reached. The strategy level is a recapitulation that retains the end result.[33]

These considerations should also provide an end to Trafford's (2017, pp. 86–88) search for *open-ended* dialogical settings: open-ended dialogical interaction, to put it bluntly, is a property of the play level. Certainly the point of the objection may be to point out either that this level is underdeveloped in the literature—a fact that we acknowledge with the provisos formulated above—, or that the dialogical approach to meaning does not manage to draw a clean distinction between local and strategic meaning—the section on *tonk* below intends to make this distinction as clear as possible.

A further point is that according to a recent paper on the dialogical interpretation of the structural rules Duthil-Novaes and French (2018, p. 147) conclude that, from the dialogical point of view, this rule is the less compelling of the structural rules.[34] The problem stems from ignoring the play level and the distinction between a local-reason *a* and the proposition *A* backed by *a*. The way to understand reflexivity, roughly, is the following: **Proponent** ! *A* (**P** claims that he has some reason for asserting *A*) **Opponent** *On what grounds?* If **O** asserted *A* before and brought forward, say, *a* as local reason, then **P** wins by responding: *my reason is the same as yours*! At the strategy level, the recapitulation of the moves leading to victory, we have **O** *a: A* and **P** *a: A*. But this is not what the play was about!

At this point of the discussion we can say that the role of the (built-in) Opponent in Lorenzen and Lorenz' dialogical logic has been fully misunderstood. Indeed, the role of *both interlocutors* (implicit or not) is not about assuring logical truth by checking the non-defeasibility of the demonstration at stake, but their role is about implementing both the dialogical definiteness of the expressions involved and the internalization of meaning.[35]

---

[33] See Rahman and Iqbal (2018).

[34] Notice however, that Duthil-Novaes and French (2018, pp. 138) seem to assume reflexivity when they bring up an example for the transitivity of implication.

[35] Notice that if the role of the Opponent in adversial dialogues is reduced to checking the achievement of logical truth, one would wonder what the role of the Opponent might be in more cooperation-featured dialogues: A *soft* interlocutor ready to accept weak arguments?

2.4.3 Pathological Cases and the Neglect of the Play Level

The notorious case of Prior's (1960) *tonk* has been several times addressed as a counterargument to inferentialism and also to the "indoor-perspective" of the dialogical framework. This also seems to constitute the background of how Trafford (2017, p. 86) for instance reproduces the circularity objection against the dialogical approach to logical constants. At this point of the discussion, Trafford (2017, pp. 86–88) is clearly aware of the distinction between the rules for local meaning and the rules of the strategy level, though he points out that the local meaning is vitiated by the strategic notion of justification. This is rather surprising as Rahman and Keiff (2005), Rahmanet al. (2009), Rahman (2012) and Redmond and Rahman (2016) have shown it is precisely the case of *tonk* that provides a definitive answer to the issue.

In this respect, three well distinguished levels of meaning are respectively determined by specific rules:

- the local meaning of an expression establishes how a statement involving such an expression is to be attacked and defended (through the particle rules);
- the global meaning of an expression results from structural rules prescribing how to develop a play having this expression for thesis;
- the strategy rules (for **P**) determine what options **P** must consider in order to show that he does have a method for winning whatever **O** may do—in accordance with the local and structural rules.

It can in a quite straightforward fashion be shown (see below) that an inferential formulation of rules for *tonk* correspond to *strategic* rules that *cannot be constituted* by the formulation of *particle* rules. The player-independence of the particle rules—responsible for the branches at the strategy level—do not yield the strategic rules that the inferential rules for *tonk* are purported to prescribe.

For short, the dialogical take on *tonk* shows precisely how distinguishing rules of local meaning from strategic rules makes the dialogical framework immune to *tonk*. As this distinction is central to the dialogical framework and illustrates the key feature of player-independence of particle rules, we will now develop the argument; we will then be able to contrast this pathological *tonk* case to another case, that of the black-bullet operator.

The *tonk* challenge and player-independence of local meaning

To show how the dialogical framework is immune to *tonk* through the importance and priority it gives to the play level, winning strategies are linked to semantic tableaux. According to the dialogical perspective, if tableaux rules (or any other inference system for that matter) are conceived as describing the core of strategic rules for **P**, then the tableaux rules should be justified by the play level, and not the other way round: the *tonk* case clearly shows that contravening this order yields pathological situations. We will here only need conjunction and disjunction for dealing with *tonk*.[36]

---

[36]Clerbout (2014a, b) worked out the most thorough method for linking winning strategies and tableaux.

A systematic description of the winning strategies available for **P** in the context of the possible choices of **O** can be obtained from the following considerations: if **P** is to win against any choice of **O**, we will have to consider two main different dialogical situations, namely those

(a) in which **O** has uttered a complex formula, and those
(b) in which **P** has uttered a complex formula.

We call these main situations the **O**-cases and the **P**-cases, respectively. In both of these situations another distinction has to be examined:

(i) **P** wins by *choosing*

- i.1. between two possible challenges *in the **O**-cases* (a), or
- i.2. between two possible defences *in the **P**-cases* (b),

iff he can win with *at least one* of his choices.

(ii) When **O** can *choose*

- ii.1. between two possible defences *in the **O**-cases* (a), or
- ii.2. between two possible challenges *in the **P**-cases* (b),

P wins iff he can win *irrespective* of **O**'s choices.

The description of the available strategies will yield a version of the semantic tableaux of Beth that became popular after the landmark work on semantic-trees by Raymond Smullyan (1968), where **O** stands for **T** (left-side) and **P** for **F** (right-side), and where situations of type ii (and not of type i) will lead to a branching-rule.

<u>Semantic tableaux and P-winning strategies for conjunction and disjunction</u>

(**P**)-Chooses	(**O**)-Chooses
$\dfrac{(\mathbf{P})\, A \vee B}{\begin{array}{c}\langle \mathbf{O}?\rangle\, (\mathbf{P})A \\ \langle \mathbf{O}?\rangle\, (\mathbf{P})B\end{array}}$	$\dfrac{(\mathbf{P})A \wedge B}{\langle \mathbf{O}?\wedge_1\rangle(\mathbf{P})A \mid \langle \mathbf{P}?\wedge_2\rangle(\mathbf{P})B}$
The expressions of the form ⟨**X** ... ⟩ constitute interrogative utterances.	The expressions of the form ⟨**X** ... ⟩ constitute interrogative utterances.
$\dfrac{(\mathbf{O})A \wedge B}{\begin{array}{c}\langle \mathbf{P}\ ?\wedge_1\rangle \\ (\mathbf{O})A \\ \langle \mathbf{P}?\wedge_2\rangle \\ (\mathbf{O})B\end{array}}$	$\dfrac{(\mathbf{O})A \vee B}{\begin{array}{c}\langle \mathbf{P}?\rangle \\ (\mathbf{O})A \mid (\mathbf{O})B\end{array}}$

However, as mentioned above, semantic tableaux are not dialogues. The main point is that dialogues are built bottom up, from local to global meaning, and from global meaning to validity. This establishes the priority of the play level over the winning strategy level. From the dialogical point of view, Prior's original *tonk* contravenes this priority.

Let us indeed temporarily assume that we can start not by laying down the local meaning of *tonk*, but by specifying how a winning strategy for *tonk* would look like with the help of **T**(left)-side and **F**(right)-side tableaux-rules (or sequent-calculus) for logical constants; in other words, let us assume that the tableaux-rules are necessary and sufficient to set the meaning of *tonk*.

Prior's *tonk* rules are built for half on the disjunction rules (taking up only its introduction rule), and for half on the conjunction rules (taking up only its elimination rule). This renders the following tableaux version for the undesirable tonk[37]:

$$\frac{(\mathbf{O})[\text{or } (\mathbf{T})]AtonkB}{(\mathbf{O})[(\mathbf{T})]B} \qquad \frac{(\mathbf{P})[\text{or } (\mathbf{F})]AtonkB}{(\mathbf{P})[(\mathbf{F})]A}$$

Tonk is certainly a nuisance: if we apply the cut-rule, it is possible to obtain a closed tableau for **T***A*, **F***B*, for any *A* and *B*. Moreover, there are closed tableaux for both $\{\mathbf{T}A, Atonk\neg A\}$ and $\{\mathbf{T}A, \neg(Atonk\neg A)\}$.

From the dialogical point of view, the rejection of *tonk* is linked to the fact there is no way to formulate rules for its local meaning that meet the condition of being player-independent: if we try to formulate rules for local meaning matching the ones of the tableaux, the defence yields a different response, namely the tail of *tonk* if the defender is **O**, and the head of *tonk* if the defender is **P**:

O-tonk rule for challenge and defence

O-move	**Challenge**	**Defence**
O ! $AtonkB$	**P** $?_{tonk}$	**O** ! B

P-tonk rule for challenge and defence

P-move	**Challenge**	**Defence**
P ! $AtonkB$	**O** $?_{tonk}$	**P** ! A

The fact that we need two sets of rules for the challenge and the defence of a *tonk* move means that the rule that should provide the local meaning of *tonk* is *player-dependent*, which should not be the case.

Summing up, within the dialogical framework *tonk*-like operators are rejected because there is no way to formulate player-independent rules for its local meaning that justify the tableaux rules designed for these operators. The mere possibility of writing tableaux rules that cannot be linked to the play level rules shows that the play level rules are not vitiated by strategic rules.

This brief reflection on *tonk* should state our case for both, the importance of distinguishing the rules of the play level from those of the strategy level, and the importance of including in the rules for the local meaning the feature of *player-independence*: it is the player-independence that provides the meaning explanation of the strategic rules, not the other way round.

[37]Cf. Rahman (2012, pp. 222–224).

The black-bullet challenge and dialogue-definiteness

Trafford (2017, pp. 37–41) contests the standard inferentialist approach to the meaning of logical constants by recalling the counterexample of Stephen Read, the *black-bullet* operator. Indeed, Read (2008, 2010) introduces a different kind of pathological operator, the black-bullet •, a zero-adic operator that says of itself that it is false. Trafford (2017, p. 39 footnote 35) suggests that the objection also extends to CTT; this claim however is patently wrong, since those counterexamples would not meet the conditions for the constitution of a type.[38] Within the dialogical framework, though player-independent rules for black-bullet can be formulated (as opposed to *tonk*), they do not satisfy dialogue-definiteness.

Let us have the following tableaux rules for the black-bullet, showing that it certainly is pathological: they deliver closed tableaux for both • and ¬•:

$$\frac{(\mathbf{P})\ \bullet}{\begin{array}{c}\langle \mathbf{O}?\rangle\\ (\mathbf{P})\bullet\supset\bot\end{array}} \qquad\qquad \frac{(\mathbf{O})\ \bullet}{\begin{array}{c}\langle \mathbf{P}?\rangle\\ (\mathbf{O})\ \bullet\supset\bot\end{array}}$$

We can in this case formulate the following player-independent rules:

Black-bullet player-independent particle rules

Move	Challenge	Defence
X ! •	**Y** $?_\bullet$	**X**! •⊃⊥

The black-bullet operator seems therefore to meet the dialogical requirement of player-independent rules, and would thus have local meaning. But if it does indeed have player-independent rules, the further play on the defence (which is a negation) would require that the challenger concedes the antecedent, that is black-bullet itself:

Deploying the black-bullet challenges

Y			**X**		
	…			…	
				! •	i
$i+1$	$?_\bullet$	i		! •⊃⊥	$i+2$
$i+3$	! •	$i+2$			
			$i+3$	$?_\bullet$	$i+4$

[38] Klev (2017, p. 12 footnote 7) points out that the introduction rule of such kind of operator fails to be meaning-giving because the postulated canonical set occurs negatively in its premiss, and that the restriction avoiding such kind of operators have been already formulated by Martin-Löf (1971, pp. 182–183), and by Dybjer (1994).

Obviously, this play sequence can be carried out indefinitely, regardless of which player initially states black-bullet. So the apparently acceptable player-independent rules for playing black-bullet would contravene dialogue-definiteness; and the only way of keeping dialogue-definiteness would be to give up player-independence![39]

2.4.4 *Conclusion: The Meaning of Expressions Comes from the Play Level*

The two pathological cases we have discussed, the *tonk* and the black-bullet operators, stress the difference between the play level and the strategy level and how the meaning provided by rules at the strategy level does not carry to the local meaning. Thus, from the dialogical point of view, the rules determining the meaning of any expression are to be rooted at the play level, and at this level *what is to be admitted and rejected as a meaningful expression amounts to the formulation of a player-independent rule, that prescribe the constitution of a dialogue-definite proposition (where that expression occurs as a main operator).*

Notice that if we include material dialogues the distinction between logical operators and non-logical operators is not important any more. If we enrich the dialogical framework with the CTT-language, this feature comes more prominently to the fore. What the dialogical framework adds to the CTT framework is, as pointed out by Martin-Löf (2017a, b), to set a pragmatic layer where normativity finds its natural place. Let us now discuss the notion of normativity.

2.5 Normativity and the Dialogical Framework

2.5.1 *A New Venue for the Interface Pragmatics-Semantics*

In his recent book, Peregrin (2014) marshals the distinction between the play level and the strategy level (that he calls *tactics*) in order to offer another insight, more general, into the issue of normativity mentioned at that start of our volume (Indeed, Peregrin understands the normativity of logic not in the sense of a prescription on *how to reason*, but rather as *providing the material by the means of which* we reason.

> It follows from the conclusion of the previous section that the rules of logic cannot be seen as tactical rules dictating feasible strategies of a game; they are the rules constitutive of the game as such. (MP does not tell us how to handle implication efficiently, but rather what

[39] We could provide at the local level of meaning a set of player-independent rules, and add some special structural rule in order to force dialogue-definiteness—see Rahman (2012, p. 225); however, such kinds of rules would produce a mismatch in the formation of black-bullet: the formulation of the particle rule would have to assume that black-bullet is an operator, but the structural rule would have to assume it is an elementary proposition.

> implication is.) This is a crucial point, because it is often taken for granted that the rules of logic tell us how to reason precisely in the tactical sense of the word. But what I maintain is that this is wrong, the rules do not tell us how to reason, they provide us with things with which, or in terms of which, to reason (Peregrin 2014, pp. 228–229).

Peregrin endorses at this point the dialogical distinction between rules for plays and rules for strategies. In this regard, the prescriptions for developing a *play* provide the *material* for reasoning, that is, the material allowing a play to be developed, and without which there would not even be a play; whereas the prescriptions of the *tactical* level (to use his terminology) prescribe how to win, or how to develop a winning-strategy:

> This brings us back to our frequently invoked analogy between language and chess. There are two kinds of rules of chess: first, there are rules of the kind that a bishop can move only diagonally and that the king and a rook can castle only when neither of the pieces have previously been moved. These are the rules constitutive of chess; were we not to follow them, we have seen (Section 5.5) we would not be playing chess. In contrast to these, there are tactical rules telling us what to do to increase our chance of winning, rules advising us, e.g., not to exchange a rook for a bishop or to embattle the king by castling. Were we not to follow them, we would still be playing chess, but with little likelihood of winning (Peregrin 2014, pp. 228–229).

This observation of Peregrin plus his criticism on the standard approach to the dialogical framework, according to which this framework would only focus on *logical constants* (Peregrin 2014, pp. 100, 106)—a criticism shared by many others since (Hintikka 1973, pp. 77–82)—naturally leads to the main subject of our book, namely immanent reasoning, or linking CTT with the dialogical framework.

The criticism according to which the focus would be on logical constants and not on the meaning of other expressions does indeed fall to some extent on the standard dialogical framework, as little studies have been carried out on material dialogues in this basic framework[40]; but the enriched CTT language in material dialogues deals with this shortcoming.

Yet this criticism seems to dovetail this other criticism, summoned by Martin-Löf as starting point in his Oslo lecture:

> I shall take up criticism of logic from another direction, namely the criticism that you may phrase by saying that traditional logic doesn't pay sufficient attention to the social character of language (Martin-Löf 2017a, p. 1).

The focus on the social character of language not only takes logical constants into account, of course, but it also considers other expressions such as elementary propositions or questions, as well as the acts bringing these expressions forward in a dialogical interaction, like statements, requests, challenges, or defences—to take examples from the dialogical framework—and how these acts made by persons

[40]This kind of criticism does not seem to have been aware of (Lorenz 1970, 2009, 2010, 2011), carrying out a thorough discussion on predication from a dialogical perspective, which discusses the interaction between perceptual and conceptual knowledge. However, perhaps it is fair to say that this philosophical work has not been integrated into the dialogical logic—we will come back to this subject below.

intertwine and call for—or put out of order—other specific responses by that person or by others. In this regard, the social character of language is put at the core of immanent reasoning through the normativity present in dialogues: normativity involves, within immanent reasoning, rules of interaction which allow us to consider assertions as the result of having intertwined rights and duties (or permissions and obligations). This central normative dimension of the dialogical framework at large, which stems from questioning what is actually being done when implementing the rules of this very framework, entails that objections according to which the focus would be only on logical constants will always be, from the dialogical perspective, slightly off the mark.

As mentioned in the introduction, in his Oslo and Stockholm lectures, Martin-Löf (2017a, b) delves in the structure of the deontic and epistemic layers of statements within his view on dialogical logic. In order to approach this normative aspect which pervades logic up to its technical parts, let us discuss more thoroughly the following extracts of "Assertion and Request"[41]:

> [...] we have this distinction, which I just mentioned, between, on the one hand, the social character of language, and on the other side, the non-social [...] view of language. But there is a pair of words that fits very well here, namely to speak of the monological conception of logic, or language in general, versus a dialogical one. And here I am showing some special respect for Lorenzen, who is the one who introduced the very term dialogical logic.
>
> The first time I was confronted with something of this sort was when reading Aarne Ranta's book Type-Theoretical Grammar in (1994). Ranta there gave two examples, which I will show immediately. The first example is in propositional logic, and moreover, we take it to be constructive propositional logic, because that does matter here, since the rule that I am going to show is valid constructively, but not valid classically. Suppose that someone claims a disjunction to be true, asserts, or judges, a disjunction to be true. Then someone else has the right to come and ask him, Is it the left disjunct or is it the right disjunct that is true? There comes an opponent here, who questions the original assertion, and I could write that in this way:
>
> $$?\vdash A \vee B\, true$$
>
> And by doing that, he obliges the original assertor to answer either that A is true that is, to assert either that A is true or that B is true, so he has a choice, and we need to have some symbol for thc choice here.
>
> $$(Dis)\frac{\vdash A \vee B\, true \quad ?\vdash A \vee B\, true}{\vdash A\, true \mid \vdash B\, true}$$
>
> Ranta's second example is from predicate logic, but it is of the same kind. Someone asserts an existence statement,
>
> $$\vdash (\exists x : A)B(x) true$$
>
> and then someone else comes and questions that
>
> $$?\vdash (\exists x : A)B(x) true$$

[41] Transcription of Martin-Löf (2017a, pp. 1–3, 7).

And in that case the original assertor is forced, which is to say, he must come up with an individual from the individual domain and also assert that the predicate B is true of that instance.

[...] So, what are the new things that we are faced with here? Well, first of all, we have a new kind of speech act, which is performed by thel oh, I haven't said that, of course I will use the standard terminology here, either speaker and hearer, or else respondent and opponent, or proponent and opponent, as Lorenzen usually says, so that's terminology but the novelty is that we have a new kind of speech act in addition to assertion.

[...] So, let's call them rules of interaction, in addition to inference rules in the usual sense, which of course remain in place as we are used to them.

[...] Now let's turn to the request mood. And then it's simplest to begin directly with the rules, because the explanation is visible directly from the rules. So, the rules that involve request are these, that if someone has made an assertion, then you may question his assertion, the opponent may question his assertion.

$$(Req1)\frac{\vdash C}{?\vdash_{may} C}$$

Now we have an example of a rule where we have a may. The other rule says that if we have the assertion $\vdash C$, and it has been challenged, then the assertor must execute his knowledge how to do C. And we saw what that amounted too in the two Ranta examples, so I will write this schematically that he will continue by asserting zero, one, or more we have two in the existential case so I will call that schematically by C0.

$$(Req2)\frac{\vdash C \quad ?\vdash C}{\vdash_{must} C'}$$

The Oslo and the Stockholm lectures of Martin-Löf (2017a, b) contain challenging and deep insights in dialogical logic, and the understanding of *defences as duties* and *challenges as rights* is indeed at the core of the deontics underlying the dialogical framework.[42] More precisely, the rules Req1 and Req2 do both, they condense the local rules of meaning, and they bring to the fore the normative feature of those rules, which additionally provides a new understanding of the notion of *implicit interlocutor*: once we make explicit the role of the interlocutor, the deontic nature of logic comes out.[43] Moreover, as Martin-Löf points out, and rightly so, they should not be called *rules of inference* but *rules of interaction.*

Accordingly, a dialogician might wish to add players **X** and **Y** to Req2, in order to stress both that the dialogical rules do not involve inference but *interaction*, and that they constitute a new approach to the action-based background underlying Lorenzen's (1955) *Operative Logik.* This would yield the following, where we substitute the horizontal bar for an arrow[44]:

[42] See Lorenz (1981, p. 120), who uses the expressions *right to attack* and *duty to defend.*

[43] This crucial insight of Martin-Löf on dialogical logic and on the deontic nature of logic seems to underly recent studies on the dialogical framework which are based on Sundholm's notion of the *implicit interlocutor*, such as Duthil Novaes (2015) and Trafford (2017).

[44] In the context of Operative Logik operations are expressed by means of arrows of the form "$\Rightarrow$".

$$\text{(Req2)}\quad \begin{array}{ccc} \vdash^{\mathbf{X}} C & & ?\ \vdash^{\mathbf{Y}}_{may}\ C \\ & \Downarrow & \\ & ?\ \vdash^{\mathbf{X}}_{must}\ C' & \end{array}$$

Such a rule does indeed condense the rules of local meaning, but it still does not express the choices while defending or challenging; yet it is the distribution of these choices that determines for example that the meaning of a disjunction is different from that of a conjunction: while in the former case (disjunction) the defender *must choose* a component, the latter (conjunction) requires of the challenger that, *her right to challenge is bounded to her duty to choose* the side to be requested (though she might further on request the other side). Hence, the rules for disjunction and conjunction (if we adapt them to Martin-Löf's rules) would be the following:

$$\text{(Dis)}\quad \begin{array}{ccc} \vdash^{\mathbf{X}} D & & ?\ \vdash^{\mathbf{Y}}_{may}\ \mathrm{D} \\ & \Downarrow & \\ & \begin{array}{c}\vdash^{\mathbf{X}}_{must}\ D' \\ \scriptstyle choose \\ \scriptstyle one\ of\ the\ components \\ \scriptstyle of\ the\ disjunction\ D\end{array} & \end{array}$$

$$\text{(Conj)}\quad \begin{array}{ccc} \vdash^{\mathbf{X}} C & & ?_{left}\ \vdash^{\mathbf{Y}}_{may}\ \mathrm{C} \\ & \Downarrow & \\ & \begin{array}{c}\vdash^{\mathbf{X}}_{must}\ C' \\ \scriptstyle assert \\ \scriptstyle the\ left\ of\ C\end{array} & \end{array} \qquad \text{(Conj)}\quad \begin{array}{ccc} \vdash^{\mathbf{X}} C & & ?_{right}\ \vdash^{\mathbf{Y}}_{may}\ \mathrm{C} \\ & \Downarrow & \\ & \begin{array}{c}?\ \vdash^{\mathbf{X}}_{must}\ C'' \\ \scriptstyle assert \\ \scriptstyle the\ right\ of\ C\end{array} & \end{array}$$

These rules can be considered as inserting in the rules the back and forth movement described by Martin-Löf (2017a, p. 8) with the following diagram:

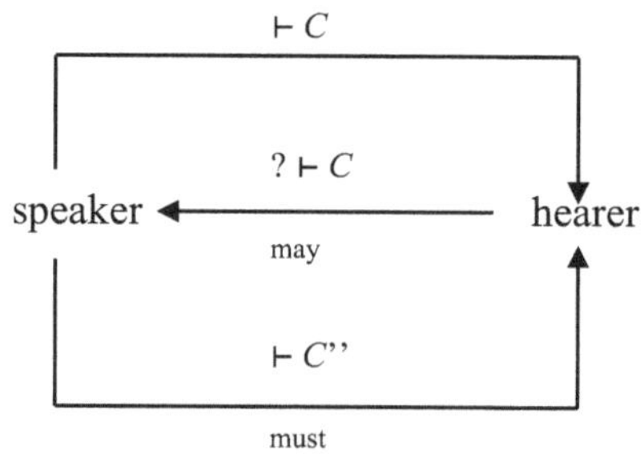

Notice however that these rules only determine the local meaning of disjunction and conjunction, not their global meaning. For example, while classical and constructive disjunction share the same rules of local meaning, they differ at the global level of meaning: in a classical disjunction the defender may come back on the choice he made for defending his disjunction, though in a constructive disjunction this is not allowed, once a player has made a choice he must live with it.

What is more, these rules are not rules of inferences (for example rules of introduction and elimination): they become rules of inference only when we focus on the choices **P** must take into consideration in order to claim that he has a winning strategy for the thesis. Indeed, as mentioned at the start of the present chapter strategy rules (for **P**) determine what options **P** must consider in order to show that he has a method for winning whatever **O** does, in accordance with the rules of local and global meaning.

The introduction rules on the one hand establish what **P** has to bring forward in order to assert it, when **O** challenges it. Thus in the case of a disjunction, **P** must *choose* and *assert* one of the two components. So, **P**'s obligation lies in the fact that he must choose, and so **P**'s *duty to choose* yields the introduction rule. Compare this with the conjunction where it is the *challenger* who has the *right to choose* (and who does not assert but request his choice). But in both cases, defending a disjunction and defending a conjunction, only one conclusion will be produced, not two: in the case of a conjunction, the challenger will ask one after the other (recall that it is an interaction taking place within a dialogue where each step alternates between moves of each of the players).

The elimination rules on the other hand prescribe what moves **O** must consider when she asserted the proposition at stake. So if **O** asserted a disjunction, **P** must be able to win whatever the choices of **O** be.

The case of the universal quantifier adds the *interdependence of choices* triggered by the *may*-moves and the *must*-moves: if the thesis is a universal quantifier of the form $(\forall x : A)\, B(x)$, **P** must assert $B(a)$, for *whatever* a **O** may choose from the domain A: this is what correspond to the introduction rule. If it is **O** who asserted the universal quantifier, and if she also conceded that, $a : A$, then **P** may challenge the quantifier by choosing $a : A$, *and request* of **O** that she asserts $B(a)$; this is how the elimination rules for the universal quantifier are introduced in the dialogical framework.

These distinctions can be made explicit if we enrich the first-order language of standard dialogical logic with expressions inspired by CTT. The first task is to introduce statements of the form " $p : A$", as we did already in the *section on local meaning of immanent reasoning*. On the right-hand side of the colon is the proposition A, on the left-hand side is the *local reason* p brought forward to back the proposition *during a play*. The local reason is therefore *local* if the force of the assertion is limited to the level of plays. But when the assertion "$p : A$" is backed by a *winning strategy*, the judgement asserted draws its justification precisely from that strategy, thus endowing p with the status of a strategic reason that, in the most general cases, encodes an arbitrary choice of **O**.

The rock bottom of the dialogical approach is still the play level notion of dialogue-definiteness of the proposition, namely

> For an expression to count as a proposition A there must exist an individual play about the statement $\boldsymbol{X}\,!A$, in the course of which $\boldsymbol{X}$ is committed to bring forward a local reason to

back that proposition, and the play reaches a final position with either win or loss after a finite number of moves according to definite local and structural rules.

The deontic feature of logic is here built directly within the dialogical concept of statements about a proposition. More generally, the point is that, as observed by Martin-Löf (2017a, p. 9), according to the dialogical conception, logic belongs to the area of ethics.

One way of explaining how this important aspect has been overseen or misunderstood might be that the usual approaches to the layers underlying logic got the order of priority between the deontic notions and the epistemic notions the wrong way round.[45]

Martin-Löf's lectures propose a fine analysis of the inner and outer structure of the statements of logic from the point of view of speech-act theory, that put the order of priority mentioned above right; in doing so it pushes forward one of the most cherished tenets of the dialogical framework, namely that *logic has its roots in ethics.*

In fact, Martin-Löf's insights on dialogical logic as re-establishing the historical links of ethics and logic provides a clear answer to Wilfried Hodges's (2008)[46] sceptical view in his Sect. 2.2 as to what the dialogical framework's contribution is. Hodges's criticism seems to target the *mathematical* interest of a dialogical conception of logic, rather than a philosophical interest which does not seem to attract much of his interest.

In lieu of a general plaidoyer for the dialogical framework's philosophical contribution to the foundations of logic and mathematics, which would bring us too far, let us highlight these three points which result from the above discussions:

(1) the dialogical interpretation of epistemic assumptions offers a sound venue for the development of inference-based foundations of logic;
(2) the dialogical take on the interaction of epistemic and deontic notions in logic, as well as the specification of the play level's role, display new ways of implementing the interface pragmatics-semantics within logic.
(3) the introduction of *knowing how* into the realm of logic is of great import (Martin-Löf 2017a, b).

Obviously, formal semantics in the Tarski-style is blind to the first point, misunderstands the nature of the interface involved in the second, and ignores the third.

2.5.2 The Semantic and Communicative Interface in Dialogical Setting

The book *Logic, Language and Method. On Polarities in Human Experience*, published in 2010, includes papers written by Kuno Lorenz in a period extending over

---

[45]See (Martin-Löf 2017b, p. 9).

[46]See also Hodges (2001) and Trafford (2017, pp. 87–88).

more than thirty years. These papers have planted the seeds for his further penetrating work (Lorenz 2009, 2010, 2011), which can be considered as philosophical variations on *Das Dialogische Prinzip* (2011, pp. 509–520) underlying what is often known as *Dialogical Constructivism*.

In the framework of Dialogical Constructivism, the analysis of the notion of intersubjectivity starts by the study of a situation where two persons are engaged in the process of acquiring a common action-competence in a situation of teaching and learning[47]; what is at stake then is not simply mirroring an individual competence in another individual, but rather it is a *procedure* which incorporates from the very beginning this dialogical situation.[48] Immanent reasoning, being an offspring of Dialogical Constructivism, inherits its philosophical background and sensitivity. In this regard, the rules of the play level are not actualizations by themselves, but are rather procedures for actualizing some action consisting in dealing with an object or appropriating that object, be it in a situation of teaching and learning, or any dialogical situation.

A consequence of Lorenz's (2011, pp. 509–520) general dialogical principle is that the interface semantics-pragmatics should be understood

1. neither as the result of the *semantization of pragmatics*—where deontic, epistemic, ontological, and temporal modalities become truth-functional operators;
2. nor as the result of the *pragmatization of semantics*—where a propositional kernel, when put into use, is complemented by moods yielding assertions, questions, commands and so on.

Lorenz's view (2010, pp. 71–79) is that the differentiation of semantic and pragmatic layers is the result of the articulation within one and the same utterance: each utterance displays in principle both features: it *signifies* (semantic layer) and it *communicates* (pragmatic layer).

Take for example one-word sentences such as:

Rabbit!
Water!

With these utterances the speaker is conveying at the same time *what* the object is and *how* the object is. But while the first aspect (*what*) is related to *object-constitution*, the second (*how*) is related to *object-description;* or, if we use the terminology of Wittgenstein's *Tractatus*, the first aspect relates to the *act of showing* and the second to the *act of saying*. Object-description is carried out by the use of predicates on an already constituted domain of objects. Lorenz recalls here Plato's *Cratylus* (388b), in which these two acts and their interdependence are distinguished as *naming* (which has the role of *indicating*) and *establishing* (with the role of *communicating*). Lorenz's

[47]The bibliographic background of this section is based mainly on (Lorenz 2010, pp. 207–218) chapter *Procedural Principles of the Erlangen School. On the Interrelation between the principles of method, of dialogue, and of reason.*

[48]The act of executing must be distinguished from taking the action as an object: while executing an action, actor and execution are said to be indistinguishable.

view is that each utterance of a sentence has this double nature, not only one-word sentences. Thus:

(a) *Sam is smoking.*

has both roles, indicating as well as communicating; though according to this analysis, uttering such a sentence does not yield any ambiguity: uttering it simply displays within one movement object-constitution (or construction) and object-description (or attribution).

While the first, object-constitution, involves differentiating parts of a whole (including the processes of partitioning a whole by synthesis and analysis), the second, object-description, involves stating that a certain relation holds. In this regard, *attribution* is not a relation, but a means for stating that relations hold of objects. The usual procedure for representing attribution by using extensional class-membership relations thus blurs this distinction.

According to the language of immanent reasoning (borrowed from CTT),

(a) *Sam is smoking*

can be read as expressing either

(b) $! \, Sam : Smoking$

or

(c) $! \, d(Sam) : Smoking(Sam) \; (Sam : Human)$

From the point of view of Lorenz's Dialogical Constructivism, me might say that the colon in both claims separates, using his words, the significative, particular, part of the expression from the communicative, universal, side of it, placed at the right side of the colon.

These considerations deserve further investigation, though this conclusion is not their place. But the point here is to stress that, according to the dialogical principle, pragmatization and semantization are two different aspects: $a : B$ and $B(a)$are not the result of an ambiguity of some sort, but are simply two aspects, the semantic aspect and the pragmatic aspect.

2.6 Final Remarks

The play level is the level where meaning is forged: it provides the material with which we reason.[49] It reduces neither to the (singular) performances that actualize the interaction-types of the play level, nor to the "tactics" for the constitution of the schema that yields a winning strategy.

---

[49]To use Peregrin's (2014, pp. 228–229) words.

We call our dialogues involving rational argumentation *dialogues for immanent reasoning* precisely because *reasons* backing a statement, that are now *explicit* denizens of the object-language of plays, are *internal* to the development of the dialogical interaction itself.

More generally, the emergence of concepts, so we claim, are not only games of giving and asking for reasons (games involving *Why*-questions) they are also games that include moves establishing *how is it that the reason brought forward accomplishes the explicative task.* Dialogues for immanent reasoning are dialogical games of *Why* and *How*. Notice that the notion of dialogue-definiteness is not bound to knowing how *to win*—this is rather a feature that characterizes winning strategies; to master meaning of an implication, within the dialogical framework, amounts rather to *know how to* develop an *actual play for it*. In this context it is worth mentioning that during the Stockholm and Oslo talks on dialogical logic, Martin-Löf (2017a, b) points out that one of the hallmarks of the dialogical approach is the notion of *execution*, which—as mentioned in the preface—is close to the requirement *of bringing forward a suitable equality while performing an actual play*. Indeed, from the dialogical point of view, an equality statement comes out as an answer to a question on the local reason *b* of the form *how*: *How do you show the efficiency of b as providing a reason for A*? In this sense the *how*-question presupposes that *b* has been brought-forward as an answer to a *why* question: *Why does A hold*? Thus, equalities express the way how to execute or carry out the actions encoded by the local reason; however, the actualization of a play-schema *does not require the ability of knowing how to win a play*. Thus, while execution, or *performance*, is indeed important the backbone of the framework lies in the *dialogue-definiteness notion of a play*.

The point of the preceding paragraph is that though actualizing and schematizing are processes at the heart of the dialogical construction of meaning, they should not be understood as performing two separate actions: through these actions we acquire the competence that is associated to the meaning of an expression by *learning* to play both, *the active* and *the passive* role. This feature of Dialogical Constructivism stems from Herder's view[50] that the cultural process is a process of education, in which teaching and learning always occur together: dialogues display this double nature of the cultural process in which concepts emerge from a complex interplay of *why* and *how* questions. In this sense, as pointed out by Lorenz (2010, pp. 140–147) the dialogical teaching-learning situation is where *competition,* the I-perspective, and *cooperation interact,* the You-perspective: both intertwine in collective forms of dialogical interaction that take place at the play level.

If the reader allows us to condense our proposal once more, we might say that the perspective we are trying to bring to the fore is rooted in the intimate conviction that meaning and knowledge are something we do together; our perspective is thus an invitation to participate in the open-ended dialogue that is the human pursuit of knowledge and collective understanding, since philosophy's endeavour is immanent to the kind of dialogical interaction that makes reason happen.

---

[50]See Herder (1960[1772], Part II).

Acknowledgements I would like to thank the Laboratory STL: UMR-CNRS 8163 and to Leone Gazziero (STL), Laurent Cesalli (Genève), and Claudio Majolino (STL), leaders of the ANR-Project *SEMAINO*, for fostering the research leading to the present study.
Many thanks to Christina Weiss for her superb editorial work. My thanks also to Zoe McConaughey (STL), Steephen Eckoubili (STL), Clément Lion (STL), Muhammad Iqbal (STL) and Mohammad Shafiei (U. Shahid Beheshti) for fruitful discussions, and to the reviewers who suggested important improvements.

References

Austin, J. L. (1946). Other minds. *The Aristotelian Society Supplementary, 20,* 148–187.
Brandom, R. (1994). *Making it explicit.* Cambridge: Harvard University Press.
Brandom, R. (1997). A study guide. In W. Sellars (Ed.), *Empiricism and the philosophy of mind* (pp. 119–189). Cambridge-Mass: Harvard University Press.
Brandom, R. (2000). *Articulating reasons.* Cambridge: Harvard University Press.
Clerbout, N. (2014a). First-order dialogical games and tableaux. *Journal of Philosophical Logic, 43*(4), 785–801.
Clerbout, N. (2014b). *Étude sur quelques sémantiques dialogiques: Concepts fondamentaux et éléments de métathéorie.* London: College Publications.
Clerbout, N., & Rahman, S. (2015). *Linking game-theoretical approaches with constructive type theory: Dialogical strategies as CTT-demonstrations.* Dordrecht: Springer.
Crubellier, M. (2014). *Aristote, Premiers analytiques. Traduction, introduction et commentaire.* Garnier-Flammarion.
Duthil-Novaes, C. (2007). *Formalizing Medieval Logical Theories.* Dordrecht: Springer.
Duthil Novaes, C. (2015). A dialogical, multiagent account of the normativity of logic. *Dialectica, 69*(4), 587–609.
Duthil Novaes, C., & French, R. (2018). A dialogical, multiagent account of the normativity of logic. *Philosophical Issues., 28*(4), 129–158.
Dybjer, P. (1994, July). Inductive families. Form Asp Comput vol. 6, pp. 440–465. *Formal Aspects of Computing, 6*, pp. 440–465.
Herder, J. G. (1960[1772]). Abhandulung über der Ursprung der Sprache. In E. Heintel, *Johann Gottfried Herder. Spachphilosophische Schriften.* (pp. 3–87). Hamburg: Felix Meiner.
Hintikka, J. (1973). *Logic, language-games and information: Kantian themes in the philosophy of logic.* Oxford: Clarendon Press.
Hintikka, J. (1996). *Ludwig Wittgenstein: Half truths and one-and-a-half truths.* Dordrecht: Kluwer.
Hodges, W. (2001). Dialogue foundations: A sceptical look. *Aristotelian Society Supplementary, 75*(1), 17–32.
Hodges, W. (2008). Logic and games. *Stanford Encyclopedia of Philosophy.*
Keiff, L. (2004). Heuristique formelle et logiques modales non normales. *Philosophia Scientiae, 8*(2), 39–59.
Keiff, L. (2007). *Le Pluralisme dialogique: Approches dynamiques de l'argumentation formelle.* Lille: PhD.
Keiff, L. (2009). *Dialogical Logic.* (E. N. Zalta, Ed.) Retrieved from The Stanford Encyclopedia of Philosophy: http://plato.stanford.edu/entries/logic-dialogical.
Klev, A. (2017). The Justification of Identity Elimination in Martin-Löf's Type Theory. *Topoi*, pp. 1-25.
Krabbe, E. C. (1985). Formal Systems of Dialogue Rules. *Synthese, 63,* 295–328.
Lorenz, K. (1970). *Elemente der Sprachkritik. Eine Alternative zum Dogmatismus und Skeptizismus in der Analytischen Philosophie.* Frankfurt: Suhrkamp.

Lorenz, K. (1981). Dialogical logic. In W. Marciszewsku (Ed.), *Dictionary of logic as applied in the study of language* (pp. 117–125). The Hague: Martinus Nijhoff.

Lorenz, K. (2001). Basic objectives of dialogue logic in historical perspective. *Synthese, 127*(1–2), 225–263.

Lorenz, K. (2009). *Dialogischer Konstruktivismus*. Berlin and New York: De Gruyter.

Lorenz, K. (2010). *Logic, Language and Method*. Berlin and New York: De Gruyter

Lorenz, K. (2011). *Philosophische Variationen: Gesammelte Aufsätze unter Einschluss gemeinsam mit Jürgen Mittelstrass greschrievener Arbeiten zu Platon und Leibniz*. Berlin / New York: De Gruyter.

Lorenzen, P. (1955). *Einführung in die operative Logik und Mathematik*. Berlin: Springer.

Marion, M. (2006). Hintikka on Wittgenstein: from language games to game semantics. In T. Aho & A.-V. Pietarinen (Eds.), *Truth and games: Essays in honour of Gabriel Sandu* (pp. 237–256). Helsinki: *Acta Philosophica Fennica*.

Marion, M. (2009). Why play logical games? In O. Majer, A. V. Pietarinen, & T. Tulenheimo (Eds.), *Logic and games: unifying logic* (pp. 3–26). Dordrecht: Springer.

Marion, M. (2010). Between saying and doing: From Lorenzen to Brandom and back. In P. E. Bour, M. Rebuschi, & L. Rollet (Eds.), *Constructions: Essays in honour of Gerhard Heinzmann* (pp. 489–497). London: College Publications.

Marion, M., & Rückert, H. (2015). Aristotle on universal quantification: a study from the perspective of game semantics. *History and Philosophy of Logic, 37*(3), 201–209.

Martin-Löf, P. (1971). Hauptsatz for the intuitionistic theory of iterated inductive definitions. In J. E. Fenstad (Ed.), *Proceedings of the Second Scandinavian Logic Symposium* (pp. 179–216). Amsterdam: North-Holland.

Martin-Löf, P. (1984). *Intuitionistic type theory. Notes by Giovanni Sambin of a Series of Lectures given in Padua, June 1980*. Naples: Bibliopolis.

Martin-Löf, P. (2014). Truth of empirical propositions: *Lecture held at the University of Leiden, February 2014*. Transcription by Amsten Klev.

Martin-Löf, P. (2015). Is logic part of normative ethics? *Lecture held at the research Unity Sciences, Normes, Décisions (FRE 3593), Paris, May 2015*. Transcription by Amsten Klev.

Martin-Löf, P. (2017a). Assertion and request. *Lecture held at Oslo, 2017*. Transcription by Ansten Klev.

Martin-Löf, P. (2017b). Assertion and request. *Lecture held at Stockholm*. Transcription by Ansten Klev.

McDowell, J. (2009). *Having the world in view: Essays on Kant, Hegel, and Sellars*. Harvard: Harvard University Press.

Peregrin, J. (2014). *Inferentialism: Why rules matter*. New York: Plagrave MacMillan.

Prior, A. (1960). The Runabout inference-ticket. *Analysis, 21*, 38–39.

Rahman, S. (2012). Negation in the logic of first degree entailment and tonk: A dialogical study. In Rahman S, Primiero, & M. Marion (Eds.), *The realism-antirealism debate in the age of alternative logics* (pp. 213–250). Springer: Netherlands.

Rahman, S., & Iqbal, M. (2018). Unfolding parallel reasoning in islamic jurisprudence (I). Epistemic and dialectical meaning within Abū Isḥāq al-Shīrāzī's system of co-relational inferences of the occasioning factor. *Cambridge Journal of Arabic Sciences and Philosophy, 28*, 67–132.

Rahman, S., & Keiff, L. (2005). On how to be a dialogician. In D. Vanderveken (Ed.), *Logic, thought and action* (pp. 359–408). Dordrecht: Kluwer.

Rahman, S., McConaughey, Z., Klev, A., Clerbout, N. (2018). *Immanent Reasoning. A plaidoyer for the Play-Level*. Dordrecht: Springer, in print.

Rahman, S., Clerbout, N., & Keiff, L. (2009). On dialogues and natural deduction. In G. Primiero & S. Rahman (Eds.), *Acts of knowledge: History, philosophy and logic: Essays dedicated to Göran Sundholm* (pp. 301–336). London: College Publications.

Rahman, S., Redmond, J., & Clerbout, N. (2017).

Ranta, A. (1988). Propositions as games as types. *Synthese, 76*, 377–395.

Read, S. (2008). Harmony and modality. In C. Dégremont, L. Kieff, & H. Rückert (Eds.), *Dialogues, logics and other strange things: Essays in honour of Shahid Rahman* (pp. 285–303). London: College Publications.
Read, S. (2010). General elimination harmony and the meaning of the logical constants. *Journal of Philosophical Logic, 39,* 557–576.
Redmond, J., & Rahman, S. (2016). Armonía Dialógica: tonk Teoría Constructiva de Tipos y Reglas para Jugadores Anónimos. *Theoria, 31*(1), 27–53.
Sellars, W. (1991). *Science perception and reality.* Atascadero-California: Ridgeview Publishing Company.
Shafiei, M. (2017). *Intentionnalité et signification: Une approche dialogique.* Paris: PHD-thesis, Sorbonne.
Sundholm, G. (1997). Implicit epistemic aspects of constructive logic. *Journal of Logic, Language and Information, 6*(2), 191–212.
Sundholm, G. (2001). A plea for logical atavism. In O. Majer (Ed.), *The logica yearbook 2000* (pp. 151–162). Prague: Filosofía.
Sundholm, G. (2006). Semantic Values for Natural Deduction Derivations. *Synthese, 148,* 623–638.
Sundholm, G. (2012). Inference versus consequence revisited: inference, conditional, implication. *Synthese, 187,* 943–956.
Sundholm, G. (2013, December 2–3). Inference and consequence as an interpreted language.
Trafford, J. (2017). *Meaning in dialogue: An interactive approach to logic and reasoning.* Dordrecht: Springer.

Chapter 3
A Phenomenological Analysis of the Distinction Between Structural Rules and Particle Rules in Dialogical Logic

Mohammad Shafiei

Abstract As it is well-known, Husserl distinguishes between three levels of formal logic: pure morphology, consequence-logic and truth-logic. The distinction between the second level, which concerns the peculiarities of the derivation of propositions from each other, and the third level, which concerns the deduction or any kind of proof as a whole in which the truth of the premises and the consequences are at work, is very important to understand Husserl's conception of pure logic. Such a distinction remains unnoticed for the truth-functional approaches toward logic, for they define the meaning of logical connectives on the basis of truth and falsity. Therefore, they cannot recognize the peculiarities of the derivation or consequence-relation as such. My aim in this paper is to show that the dialogical distinction between particle rules and structural rules, and also the distinction between play-level and strategy-level, may be considered as representing the distinction between the second and the third levels of logic; and thus the dialogical semantics provides us with a device to explore the phenomenological idea in a precise way. On the other hand, certain themes of the overall phenomenological analysis of logic help to show the philosophical significance of those dialogical features.

Keywords Dialogical logic · Transcendental logic · Truth · Consequence · Structural rules · Particle rules · Phenomenology · Intentionality · Play-level · Strategy-level

3.1 Introduction

Formal logic, according to Husserl (1969, pp. 48–55), has three levels. The first level is the study of pure morphology of judgment, or generally, of various grammatical forms of expression which are used in reasoning. The second level is what Husserl

M. Shafiei (✉)
Shahid Beheshti University, Tehran, Iran
e-mail: m.shafiyi@gmail.com

C. Weiss (ed.), *Constructive Semantics*, Logic, Epistemology, and the Unity of Science 44, https://doi.org/10.1007/978-3-030-21313-8_3

calls consequence-logic. The third level is truth-logic.[1] The distinction between the latter two levels is very important in regard to the investigations concerning the nature of logical argumentations.

In the present note I will try to elaborate some aspects of the aforementioned distinction linking it with some ideas developed within dialogical semantics. However, there is a straightforward explanation of this distinction using model-theoretic notions. According to this explanation we can understand Husserl as distinguishing on one level an uninterpreted (or re-interpretable) formal system with its derivation rules and on another level an interpreted, full deductive system with its laws concerning truth and falsity. Such an explanation fits good to the historical background and it is quite natural concerning the foundational problems of logic which were in the air at that time. The explanation I am going to offer is not so much historically straightforward, for I will try to recall some notions of dialogical logic in order to explain Husserl's idea of logic. However, this deliberate anachronism is not without reason. I am rather concerned with the philosophical significance of Husserl's distinction as it should be understood in accordance with his overall philosophical approach, namely beside other ideas of the method of transcendental phenomenology, not only with relevance of the aforementioned distinction to the then-trending debates. In this respect it is expectable to recognize some insights which would lead or support some technical improvements which were not available at that time. The improvements I particularly point out are the distinction between play-level and strategy-level in the dialogical framework and that between particle rules and structural rules—which historically goes back to the difference between operational rules and structural rules developed by Gentzen in his sequent calculus, presented some years after the publication of *Formal and Transcendental Logic* and itself based upon an earlier work by Hertz (1922), of which Husserl apparently was not aware.

The main idea I am going to develop is to regard the distinction between consequence-level and truth-level not as one between two logical disciplines but rather as one between two different levels of study in every phase of constituting and developing a formal logic. If we regard each level as a distinct discipline it would be acceptable to regard consequence-logic as endorsing a rather traditional, and at the same time model-theoretic, conception of consequence based on the notion of contradiction (namely B is a consequence of A iff 'A and not-B' is contradictory). However, if we understand the distinction as phaseal, then such an interpretation isn't appropriate any more. In addition, new horizons more apt to the transcendental attitude can thus be opened up.

First of all, I attempt to put the distinction between truth-level and consequence-level in the phenomenological background by considering three important ideas of this method. In the next part I will argue that such a distinction finds its adequate realization in the dialogical framework. I will introduce a short description of dialogical semantics, then explain why such a framework is more in accordance with transcendental phenomenology than the model-theoretic and also proof-theoretic frameworks

[1] A helpful explanation of these distinctions has been given by Bachelard in (1968, pp. 11–24). See also the section "Husserl's Idea of Pure Logic" in (Shafiei 2018).

with respect to the mentioned three phenomenological ideas and consequently the distinction in question.

3.2 Some Phenomenological Considerations

In the first chapter of part I of *Formal and Transcendental Logic*, Husserl distinguishes between consequence-logic and truth-logic. He says about the former:

> It is an important insight that questions concerning consequence and inconsistency can be asked about judgments *in forma*, without involving the least inquiry into truth or falsity and therefore without ever bringing the concepts of truth and falsity, or their derivatives, into the *theme*. In view of this possibility, we distinguish a level of formal logic that we call consequence-logic or logic of non-contradiction (Husserl 1969, p. 54).

Accordingly, in a following section, he offers two different readings of the logical principles according to each level. For example, in the consequence level modus ponens reads:

> "N" follows analytically from two judgments of the form "If M, then N" and "M".

While:

> The corresponding truth-principle then reads:
>
> If an immediate relationship of total analytic antecedent and total analytic consequent obtains between any two judgments, M and N, then the truth of the antecedent entails the truth of the consequent… (Husserl 1969, p. 67).

However, the distinction should not be conceived of merely as a reduplication of principles of logic or as the difference between formal rules of inference and their application in the ontological regions. In order to grasp what the effects of the separation between truth and consequence in logical studies are we suggest to consider two basic ideas concerning truth in general and also a phenomenological observation concerning the articulation of inference and deduction.

The first point relevant here is the idea of intentionality and the distinction hereupon between the acts of intention and those of fulfillment. Accordingly, we observe that truth concerns basically the side of fulfillment, since in truth we don't have an empty intention but an experience of a, static or dynamic, coincidence of fulfillment. On the other hand, the consequence-relation, or as Husserl occasionally says the analytical includedness and excludedness, is primarily a relation of including or excluding among (propositional) intentions. Therefore, if there is a logical issue which first appears on the consequence-level, or is considered as pertaining to this level, it should first be analyzed in terms of intentions rather than in terms of properties or conditions of fulfillment.

The second point is that truth, according to transcendental phenomenology, is always an experienced truth. We are not justified in speaking of truth absolutely independent of the acts of consciousness. If we consider the indispensable role of acts of truth-experiencing, another important matter arises with regard to the distinction between the three levels of logic. Acts of truth-experiencing, either in a static or a dynamic way, have their own peculiarities among the acts of consciousness. And more importantly they deal with the peculiarities of fulfillment. According to the mentioned distinction, we should distinguish between three kinds of acts belonging to the realm of logic. On the first level we have acts of judgment-forming, namely constituting forms of judgment, that is propositions, which belong to the general sphere of acts of meaning-constitution and in particular the acts of *categorial synthesis*. On the second level, we have acts of consequence-following, which deal with intentions not in order to forming them but detecting or constructing patterns of includedness and excludedness among them. On the third level we have acts of truth-experiencing to which belong acts of proof-constructing. So these three kinds of acts are different and the distinction among the three levels of logic can also be regarded as a hint not to confuse those acts and not to reduce one kind to another.

The third point is that it is *possible* that an ideal object is objectively intuited. Indeed, there is no essential difference between the evidence of ideal objects and that of individual objects (Husserl 1969, p. 155). Ideal objects include categorial forms, namely those dependent moments which appear in categorial articulation—or synthesis thereof. Therefore, categorial objects are not merely subjective artifacts in this or that manner, but they can be genuinely objective moments which can be experienced in their own manner:

The ideal object, the ideal simple object (I do not mean that composed in categorial synthesis) and also every categorial object, that goes to ideals, is that which in the ideation, in the categorial intuition, is given or to be given completely and firmly.[2]

The objectivity of categorial forms concerns our investigation for in every kind of formal inference there are categorial articulations of the judgments which make such an inference possible. Therefore, the logical connectives that are of an indispensable role in the consequence-relation themselves can be objective. This directly affects the meaning-explanation of the logical connectives. From the phenomenological point of view logical connectives have their own objective status and cannot be reduced to some mere external relations between predetermined truths and falsities. One of the dominant presuppositions of logical studies is a kind of "atomism" which takes elementary propositions as (possibly) representing the facts while logical connectives are taken to be at use only to make some combination of the former. In such a view the logical connectives are nothing but functions over the truth or provability of the elementary terms. This view is rejected by transcendental phenomenology. Therefore, both, the truth-functional and the proof-theoretic meaning-explanation of the logical connectives are declined, for both see logical connectives as kinds of external functions without any possibility for considering them as objective moments. Indeed, the truth-functional approach explicitly, and the proof-interpretation indirectly, recall the notion of truth while speaking of logical connectives which according to their very nature first of all appear on the consequence-level. If we can offer a meaning-

[2]Der ideale Gegenstand, der ideale schlichte Gegenstand (ich meine nicht in kategorialer Synthesis gefasst) und ebenso jeder kategoriale Gegenstand, der auf Ideales geht, ist ein in der Ideation, in der kategorialen Anschauung abgeschlossen, fest, fertig Gegebenes oder zu Gebendes (Husserl 1986, p. 210).

explanation for the logical connectives on the consequence-level, we will also be able to save an objective place for them.

To summarize this section, I briefly restate the mentioned points and their link to the distinction between consequence-level and truth-level. There is in general a distinction between intention and fulfillment, which also concerns the case of judgment and argumentation. From a certain point of view, consequence-level deals with some properties of propositional intentions while on truth-level peculiarities of fulfillment are concerned. Truth is always a matter of experience. Additionally, the acts involved in attaining truth have their own peculiarity and are different from other types of acts. In relation to the distinction between consequence-level and truth-level we should distinguish between the acts of derivation and of developing a consistent theory on the one side and the acts of truth-establishing and proof-construction on the other side. Besides, the meaning of logical connectives cannot be reduced to their functionality for the truth or falsity (or provedness or lack thereof) of some atomic propositions, that is their essence, should not be placed on the truth-level only. Their meaning should be explained on the consequence-level without any reference to truth and falsity of the involved, elementary propositions.

Considering the above remarks we have some criteria at hand for a technical framework to study and develop pure logic in a way phenomenologically admissible. In the following, I will argue that the dialogical approach is quite appropriate in this respect. First I introduce in a very short way the basics of the aforementioned method.

3.3 A Short Introduction to Dialogical Semantics

The dialogical approach has been introduced by Paul Lorenzen and Kuno Lorenz and further developed by Shahid Rahman and others. It first was supposed to provide a semantics for intuitionstic logic but since then it has been demonstrated that it could provide semantic rules for a varied range of logical systems so that it is an excellent framework to represent and evaluate different logics. For a study of the various dialogical semantics for different logical systems see (Rahman and Keiff 2005). The dialogical approach provides us with a semantics which is neither model-theoretic nor proof-theoretic.[3]

In the dialogical semantics there are two parties, the proponent **P** and the opponent **O**. The proponent introduces a thesis and defends it against the attacks of the opponent. If there is a winning-strategy for the proponent with respect to a statement, that statement is valid. The attacks and responses are to be performed according to two kinds of rules, particle rules and structural rules. Particle rules are those governing local moves, namely they determine how each form of complex formulas can be attacked and defended. Structural rules are those governing the whole dialogue; they determine the rights and the obligations of each party, how a dialogue proceeds and terminates and who is the winner. Dialogical logic accordingly distinguishes

[3]For a good explanation of this point see Rückert (2001).

between the play-level, which consists in the moves of derivation according to the rules, and the strategy-level, which consists in ordering a chain of such moves in order to establish a proof if available. For an elaborate study of the various aspects of the dialogical framework see (Rahman et al. 2018).

I give a short introduction of the dialogical rules in the following.[4]

Particle Rules

For any logical connective there is a particle rule which determines how to attack and defend a formula with a specific main connective. These rules are standard in the literature:

	Attack	Response
$A \vee B$	$?_{\vee}$	A, or B (The defender chooses)
$A \wedge B$	$?_{L}$, or $?_{R}$ (The attacker chooses)	A, or B (respectively)
$A \to B$	A	B
$\neg A$	A	(No possible response)
$\forall xA$	$?_{\forall x/c}$ (The attacker chooses c)	$A[x/c]$
$\exists xA$	$?_{\exists x}$	$A[x/c]$ (The defender chooses c)

Structural Rules

Structural rules determine the structure of the interactions which form a certain argumentation. We have:

(SR-0) Starting Rule: The Proponent begins by asserting a thesis.
(SR-1) Move: The players make their moves alternately. Each move, with the exception of the starting move, is an attack or a defense.
(SR-2) Winning Rule: Player **X** wins iff it is **Y**'s turn to play and **Y** cannot perform any move.
(SR-3) No Delaying Tactics Rule: Both players can only perform moves that change the situation.[5]

[4]The following rules are standard within dialogical studies. However in the current manner of presentation, I particularly benefited from the representations given in Rebuschi (2009) and Rahman and Keiff (2005).

[5]In order to handle this rule it should be determined for each party how many times it may repeat the same attack (repetition of defense is redundant in any case and it is not allowed). This is called *repetition ranks*. Clerbout (Clerbout 2014) has shown that, in propositional logic, it would be sufficient to assign the rank n + 1 to the proponent while the rank n is assigned to the opponent, namely there is a winning strategy for a formula if and only if there is a winning-strategy for that formula while the proponent is allowed to attack once more than the opponent. Therefore, I do not specify the ranks in the following dialogues, and one can suppose that it is 1 for **O** and 2 for **P**.

(SR-4) Formal Rule: **P** cannot introduce any new atomic formula; new atomic formulas must be stated by **O** first. Atomic formulas can never be attacked.
(SR-5c) Classical Rule: In any move, each player may attack a complex formula uttered by the other player or defend him/herself against *any attack* (including those that have already been defended).
(SR-5i) Intuitionistic Rule: In any move, each player may attack a complex formula uttered by the other player or defend him/herself against *the last attack that has not yet been defended.*

I give a simple example to show how this semantics works. Let us examine the formula $p \to (q \to p)$.

Remark on notation: The moves are given in the order of utterance. The parenthesized information indicates the number of the move, whether it is an attack (a) or a response (r) and to which move it is an attack or a response. The long dash indicates that there is no further move and the player has lost.

O		**P**	
		$p \to (q \to p)$	(0)
(1a0)	p	$q \to p$	(2r1)
(3a2)	q	P	(4r3)
	…		

Here in the move 4, **P** is able to respond to the attack, because **O** has asserted p before; and since there is no other move possible for **O**, **P** wins and the formula is demonstrated to be valid.

3.4 The Distinction Between Consequence-Level and Truth-Level in the Dialogical Method

It is clear from the above description that the structural rules do not refer to the logical connectives. Their role and conditions of correctness are on a different level than those of the particle rules. On the other hand, the particle rules are independent of the structural rules and are also independent from each other.

Therefore, at first the separation between particle rules and structural rules is a realization of the distinction between consequence-level and truth-level. The two sorts of dialogical rules have their own properties which fit well to our phenomenological distinction. Particle rules are symmetric, namely they do not refer to the role of the players (whether proponent or opponent) engaging in the moves concerning a logical connective. Whereas structural rules are in general asymmetric and they are meant to determine the role of the players, namely their obligations and rights. The fact that on the truth-level we have *positing* acts and accordingly have to deal with the *position* of the asserter, is already clear from the phenomenological investigations.

So, it is an apt feature that in dialogical logic, in its structural rules, which stand on the side of truth-logic, we deal with the positions of players and the properties of their acts. On the other hand, a mere intention needn't be positing. Thus, if the particle rules are primarily meant to reflect the intentions behind logical connectives, they should not depend on the position of the asserter, hence on the symmetry of particle rules.

The discussed feature is unsurprisingly not observed in truth-functional semantics. In truth-functional semantics it is not possible to separate the meaning of a logical connective from the (alleged) truth of a logical principle in which it appears. For example the truth-functional definition of negation inherently presupposes the correctness of the law of excluded middle. T Or as another example, the definition of implication takes for granted the correctness of *ex falso*. From the phenomenological point of view this is because the truth-functional, and in general the model-theoretic, approach tries to define an intention by means of the conditions of its fulfillment. Then a commitment in regard to these conditions would be required in any case. Here the determination of a consequence-relation by appealing to the notion of truth (and falsity), which is a case of confusion between intention and its fulfillment, violates the distinction between consequence-level and truth-level. A comparison makes obvious the advantages of dialogical semantics in this respect. For example, the validity of the law of excluded middle is not presupposed in the rule concerning negation: dialogues for both intuitionistic and classical logic use the same particle rule, but the former rejects the law of excluded middle while the latter allows it. This is because of the difference in the structural rules.

The dialogical framework makes it possible, through its distinction between particle rules and structural rules as considered respectively for the consequence-level and the truth-level, to fix the meaning of the logical connectives and then study the truth involving them. Thus we are able to evaluate different logical systems and assess the principles they claim without loosing a common ground and without falling into a wholistic relativism. This latter feature is a phenomenologically required condition. In other words, in dialogical logic, we have a framework to perform the phenomenological reduction for investigating the essence of logical connectives on the consequence-level, in their purity and in separation from each other, while the judgments concerning the truth of some alleged principles in which those connectives appear are put aside. On the other hand, we can also carry out the phenomenological reduction to investigate the essential features of validity, namely that of formal truth, without confusing these essential features with mere indications of the definitions of logical connectives.

Secondly, the distinction between consequence-logic and truth-logic is reflected on in the distinction between play level and strategy level. Both, structural rules and particle rules are recalled in a dialogue in order to argue about a thesis. However, we still have a distinction between the acts constituting a dialogue and the acts considering the possible dialogues in order to find an effective strategy to demonstrate the validity or non-validity of a thesis. Again the first activity belongs to the consequence-level and the second one to the truth-level. In other words, the structural rules by themselves do not establish the validity since they can be used in a consistent argument but be fruitless with regard to truth. In this view, dialogical logic acknowledges a different type of activity from a higher stage and calls it strategy-level in contrast to the play-level which only concerns the correctness of a particular game. This can be seen as related to the distinction between three types of logical acts which we mentioned under the second point in Sect. 3.2.

In a yet another stage, a whole deductive system can be seen as belonging to the consequence-level in comparison with a discipline which applies it to the ontological regions. I repeatedly emphasize that the distinction doesn't imply that there are two distinct and mutually exclusive fields of logic called consequence-logic and truth-logic. Rather these are two levels of logical studies and appear in every phase from the beginning to the highest one.[6] So for example, we cannot say that in constituting the meaning of logical connectives there is no idea of truth at all. Rather there shouldn't be any reference to the truth of the propositions involved. On the play-level we do have a conception of truth but not of a particular truth that the play may aim at.

Above, I explored some basic accordances of dialogical logic with transcendental phenomenology with respect to the latter's distinction between three levels of logic. More fundamental accordances I have discussed in detail in (Shafiei 2018, Chap. 3). Here I wanted to show firstly how the dialogical framework is appropriate to realize the aforementioned phenomenological distinction and secondly how phenomenology approves the essential features of dialogical semantics.

Acknowledgements This paper is a revised version of the talk I gave in the workshop *Constructive Semantics*. I warmly thank the organizer, Dr. Christina Weiss, for the invitation and also the participants for the valuable comments.

References

Bachelard, S. (1968). *A study of Husserl's formal and transcendental logic* (L. E. Embree, Trans.). Evanston: Northwestern University Press.

Clerbout, N. (2014). Finiteness of plays and the dialogical problem of decidability. *IfCoLog Journal of Logics and their Applications, 1*(1), 115–130.

Hertz, P. (1922). Über Axiomensysteme für beliebige Satzsysteme. I. Teil. Sätze ersten Grades. *Mathematische Annalen, 87*, 246–269.

Husserl, E. (1969). *Formal and transcendental logic* (D. Cairns, Trans.). Netherlands: Martin Nijhoff, The Hague.

Husserl, E. (1986). *Vorlesungen über Bedeutungslehre Sommersemester 1908, volume XXVI of Husserliana*. Netherlands: Springer.

Rahman, S., & Keiff, L. (2005). On how to be a dialogician. In D. Vanderveken (Ed.), *Logic, thought and action* (pp. 359–408). Netherlands: Springer.

Rahman, S., McConaughey, Z., Klev, A., & Clerbout, N. (2018). *Immanent reasoning or equality in action*. New York: Springer.

Rückert, H. (2001). Why dialogical logic? In H. Wansing (Ed.), *Essays on non-classical logic* (pp. 165–185). River Edge: World Scientific.

[6]See for example (Husserl 1969, p. 137).

Rebuschi, M. (2009). Implicit versus explicit knowledge in dialogical logic. In O. Majer, A. V. Pietarinen, & T. Tulenheimo (Eds.), *Games: Unifying logic, language, and philosophy* (pp. 229–246). Dordrecht: Springer.

Shafiei, M. (2018). *Meaning and intentionality: A dialogical approach*. London: College Publications.

Chapter 4
A Dialogical Account of the Intersubjectivity of Intuitionism

Clément Lion

Abstract The present paper aims at integrating the phenomenological reading of Brouwerian intuitionism into the domain of semantics, by challenging the claim that the very meaning of mathematical expressions—expressions of free choice sequences included—is invariable and objectively determinable and that, accordingly, any deictic expression should be removed from mathematics. By introducing constructability into the constitution of meaning itself and by considering meaning as a "social act", we try to map another route into intersubjectivity, based on the distinction between the *play-level* and the *strategic level*, which has been further developed in the dialogical framework, following the work of Paul Lorenzen. It is suggested that the steps towards such a route can be retraced from Oskar Becker's original "Cartesian" approach to intersubjectivity, which facilitates a new reading of Brouwer's own way of conceptualizing "mutual understanding". In doing so, our general purpose is therefore to promote an insertion of dialogical constructivism into Mark van Atten's take on the intuitionist *Creating Subject*.

Keywords Actual/potential · Creating subject · Deictics · Dialogical logic · Free choice sequences · Intersubjectivity · Intuitionism · Occasional expressions · Phenomenological reduction · Transcendental subject · Turing machine

4.1 Essentially Occasional Expressions and Choice Sequences

Deictic or "essentially occasional" expressions lead to serious difficulties in semantic frameworks based on the claim that any meaning could *ideally* be expressed through objectively fully determinate expressions: it forces one to acknowledge a *fluctuation* in meaning. When words like "I", "here" or "now" are used by someone, a full understanding of their meaning is possible only by paying attention to occasional contexts in which they are used, so that a meaning, which is based on such factual

C. Lion (✉)
Université de Lille, Lille, France
e-mail: clement.lion@univ-lille.fr

C. Weiss (ed.), *Constructive Semantics*, Logic, Epistemology, and the Unity of Science 44, https://doi.org/10.1007/978-3-030-21313-8_4

contexts, cannot practically be reduced to any objective meaning. A discourse, based on such uses, is essentially fluctuant because to know its pure expression is obviously not enough to know exactly what it means: in order to understand it, attention must be paid to the concrete speaking person and to its situation, so to speak "from the inside". According to Husserl (Husserl 2001a, 218) there can be no such occasional expressions in mathematics, or in any "abstract" science, because the circumstances of actual discourse cannot have any influence on objective meaning,[1] and mathematical meaning *has to be* objective and entirely determinate. Therefore, Husserl pays attention to occasional expressions only as a *marginal* matter, which has to do with expressions of "common life" and clearly not with mathematical, well-defined abstract expressions.[2]

One could show nevertheless that there *is* actually a mathematical application of occasional expressions (for instance a mathematical use of "I"). The matter would also acquire a very different status, and it would become a crucial task to solve the problem of the full communicability of essentially occasional or *subjective* expressions. As a matter of fact, in Brouwerian intuitionism—and especially in some controversial arguments such as the so called *Creating Subject argument*—, there actually *is* such a mathematical use of *subjective* expressions. To begin with, *free choice sequences* are (controversial) mathematical objects, based on subjective intra-temporal choices, in the sense that they depend on time, and whose mathematical properties depend on the temporal context from which they are viewed. In *Mathematische Existenz*, Oskar Becker—whose original proposal was to use the methods of transcendental-constitutive phenomenology to gain insights into the dispute between Intuitionism (Brouwer) and Formalism (Hilbert)—opened the way to a phenomenological reading of such mathematical objects (Becker 1927, 504). He provided the very first basis of a specific semantic framework for constructive objects, using theoretical means furnished by Husserl in *Logical Investigations* and *Ideas pertaining to a pure Phenomenology*. The point was, among others, to establish the non-reliability of the principle of the excluded-middle for some judgments (namely those implying potential infinity) by using the theory of judgment Husserl deployed in the *Sixth Investigation* (starting from the famous distinction between *intention* and *fulfilment*). It then becomes a matter of supplying explicit philosophical bases to the implicit semantics—that is to say the particular understanding of what *meaning* is—underlying Brouwer's own theses. To take Becker's example literally, if one pays attention to the sequence which begins with the natural numbers: 2 4 25 3 6 6 7... without any knowable law, one can wonder whether the number 1 belongs to the sequence. As it is not presently the case, neither "1 belongs to the sequence" nor "1 does not belong to the sequence" is true; but it *could* later be the case that the *subject*, who is *creating* the sequence, chooses the natural number 1, so that the first

[1]"Meaning" is the translation of the *Husserlian* use of the word "*Bedeutung*" (which he distinguishes from "*gegenständliche Beziehung*" of a sign). It does not correspond to the *Fregean* use. Within Husserl's works, *Sinn* and *Bedeutung* are synonymous.

[2]This is actually also M. Dummett's point of view: "Any temporal reference we introduce into mathematical statements must be non-indexical" (Dummett 1977, 247).

proposition, which is presently neither true nor false, comes true, in the sense that the corresponding intention becomes fulfilled. If this proposition's truth or falsity is not determinate every time, then its *expression* itself has to be fluctuant, since it must depend on the way the subject uses it to express what he is presently creating and modifying. Though it isn't Becker's purpose to give an account of the way expressions of choice sequences can be given a determinate meaning, it appears that an expression such as α (=2 4 25 3 6 6 7...) has to be subject-dependent, if it has to express a contextually modifying sequence.[3] If one follows Brouwer's conception of any act of mathematical intuitionism as being ascribable to the Creating Subject (van Atten 2002a, 5), one can ask what the semantical status of the expression of such a subject (and of its own acts) can be. As it is a temporally determinate—so to say "living"—subject, whose determinations depend on a temporal (internal) context, its use of occasional expressions hardly seems to be replaceable by fully determinate expressions (because openness must be taken into account).

The point of its meaning and of its possible expression might deserve to be examined fundamentally in relation to Husserl's analysis of mutual understanding for essentially occasional *expressions*, that is in terms of possibly unambiguous expressions of specific constructive objects, like choice sequences, through real acts of speech in a determinate language. It should be noted here that Brouwer's claims about language being secondary in mathematical practice and about the impossibility of exactness through language[4] make it especially necessary to give an account of occasional expressions, that is of expressions whose meaning fluctuates according to the free-unfolding of subjective contexts.

The clearest insertion of phenomenological insights into the development of a specific semantic framework for intuitionism starts from Becker's use of the aforementioned distinction between intention and fulfilment, in order to distinguish two sorts of negations (if *intention* is defined as the act directed toward an object, then *not being fulfilled* for the moment is different from being exploded by a "frustrating" construction) (Becker 1927, 499; Husserl 2001b, 211).

We consider that the intuitionistic semantics of negation appeared first in *Mathematische Existenz*, and that the next decisive step was taken by (Heyting 1931). In

---

[3]According to (van Atten 2007, 89), a lawless choice sequence must not be confused with the meaning of its expression, i.e. with the concept under which it falls. Accordingly, the meaning of the expression "the lawless sequence begun by me exactly one year from now at 6.58 pm" is invariant through time, whereas the corresponding object itself changes. The objectivity of the meaning of such an expression presupposes however obviously the possibility to set up a *unique* time scale, established from the deictic indication "now", for any subject. To grasp the meaning of this expression, one has also certainly to pay attention to its "empractical integration", following K. Bühler's words, to refer to the way a speech depends on the concrete situation in which it is expressed (Bühler 2011, 45–47). M. van Atten refers to Husserl's concept of a "noematic nucleus", by contrast with "sense moments [...] in an extending meaning" (Husserl 1983, 218). It could be actually the case, as we suggest in the following study, that a choice sequence be, *as an object*, the very product of speech acts, as it is, incidentally, suggested by (van Atten and van Dalen 2002, 331).

[4]"Es gibt also auch für die reine Mathematik keine sichere Sprache, d.h. keine Sprache, welche in der Unterhaltung Missverständnisse ausschließt und bei der Gedächtnisunterstützung vor Fehlern (d.h. vor Verwechselung verschiedener mathematischer Entitäten schützt" (Brouwer 1975, 421).

this paper, Becker's authorship is clearly acknowledged and its scope is broadened by application to other logical connectives, like *disjunction*. The result of these works now holds as the so-called BHK semantics, in which the meaning of the different logical connectives is provided in terms of the provability of elementary propositions on which they are applied, that is in terms of the constructability of a content fulfilling the intention they express. The *Logical Investigation*'s key assumption of the ideality of meaning is preserved in constructivist interpretations, in the sense that an expression does not have to be effectively fulfilled in order to be meaningful: possible fulfilment is sufficient for meaningfulness to be determinate. Effective constructions are not part of the meaning of propositions, but of their truth or falsity. Accordingly, there is no need to identify *which* subject is carrying out the construction, nor in which "mental" context this takes place.[5] It is possible that occasional expressions constitute the "blind-spot" of constructive semantics on this precise concern. How can it be that the meaning of a proposition like "1 belongs to α" does not depend on the *effective* construction of α? Certainly, such a proposition is possibly neither true, nor false, but it seems nevertheless that its very meaning has to be determinate, before

[5]Such a point is expressed by Heyting, in a reported discussion with Kreisel, when he says that "it is irrelevant which subject makes the construction" and that the only thing that matters is to express the way "intuitionists make their constructions, without reference to the subject" (Kreisel 1967, 173). It is clear that if one wants to refer explicitly to a determinate subject or to a determinate "I" in any mathematical expression, it is no longer possible to base a determinate constructive semantics on the *Logical Investigations* as such. In a sense, one could see Kreisel's attempt to give an axiomatic reconstruction of the Creating Subject as an effort to get rid of any deictic expression in mathematics of choice sequences, and to replace subjective expression ["*he*" i.e. "the Creating Subject *now*/at *that* moment" in Brouwer's own discourse (Brouwer 1975, p. 478)] with a well-defined new kind of sentential operator. Though Heyting and Kreisel disagree on the relevance of attempts to axiomatize the Creating Subject, they agree with most constructivists on the necessity to obliterate any trace of proper *subjectivity* in rigorous renderings of the intuitionist way of reasoning (Kreisel 1967). One can also consider that Troelstra does something similar, arguing that lawless choice sequences can be spoken of "without introducing elements of *arbitrariness* (*subjectiveness*) into mathematics *where it does not belong*" (Troelstra 1977, 12). If we were able to replace the way Brouwer talks of subjective parameters for dynamical mathematical objects as free choice sequences with well-defined rules of "constructability", or at least of "verifiability", then it would confirm Husserl's assumption that the language of mathematics has nothing to do with *subjective* or *occasional* expressions, *even in the case of controversial arguments like those depending on the Creating-Subject-device*. However, if it can be shown that such a replacement is not possible—and it has been convincingly claimed it is not, at least in its axiomatic form (Sundholm 2014)—and that its irreducible subjective compounds cannot ever get any objective substitute, then it would justify going back to Husserl's treatment of occasional expressions (and to its limits) in order to see if intersubjective "transfer" of meaning can be thought in the case of deictic expressions, irreducibly to "objective" meaning, but nevertheless so that it would still count as a *possible* basis for a claim to mathematical validity. In fact, the consequence of Sundholm's result is that the arguments, which are deployed against the principle of excluded-middle, cannot have any mathematical *constructive* validity based on the Creating-Subject-device, as long as the latter is identified with its axiomatized form—then it does not disqualify Brouwer's own treatment of it, as indicated in (van Atten 2018, footnote 48). Indeed, it seems that there cannot be a constructive object corresponding to this device, because the only conceivable way to do it is to "*postulate* a non-constructive LEM-proof" (Sundholm 2014, 22). If one wants to refer to the Creating Subject in the broadest sense and to consider the totality of intuitionist mathematic as being nothing else but "an elaboration of the Creating Subject"

any effective evaluation of its provability can be carried out.[6] It is perfectly possible that the general perspective which is deployed in *Logical Investigations* leaves space for intuitionism, only on its own fringes, that is only where occasional expressions are given a specific status.

4.2 Meaning as a Social Act

In the first part of *Logical Investigations*, called "Expression and Meaning" Husserl's position about "essentially occasional expressions" is actually rather ambiguous. According to him, any "subjective expression" could "ideally" be replaced by objective expressions, though such a substitution will ever be impossible to achieve (Husserl 2001a, 223): such an "ideality" is essentially connected with the claim that *there are* "truths in themselves". If truths are determinate in themselves, then it must be possible to provide fully determinate expressions of them. This substitution being considered by Husserl as *unfeasible* in reality, the meaning of that kind of expressions must either be ascribed to an ideal being (which is then of no use for us) or remain only partially determinate. In this second case, one can ask if it is nevertheless possible to consider that not fully determinate *expressions* can transfer determinate *meaning* from one subject to another and that such an intersubjective transfer of *determinate* meaning for *indeterminate* content is the only way free choice sequences can be ever communicated.

A solution to this problem could be extracted from several distinctions provided by Husserl in *Logical Investigations*. By distinguishing "acts of meaning" and "meaning" (Husserl 2001a, 224) (the latter being thought of as the ideal unity of all possible executions of the former), he opens up the possibility for an expression to be fluctuant in its act of meaning-constitution without this being the case for meaning itself. By distinguishing "*indicative* meaning" and "*indicated* meaning", (Husserl 2001, 219) delivers a way to conceive at once determination of meaning in intersubjective transfer *and* dependency of the same meaning to the acts of a particular (talking) subject.

A crucial point can also be the following. Husserl states, in an addendum, which is located in the *Sixth Logical Investigation*, that, in the case of occasional expressions, meaning is not constituted in the same way by the speaker and the hearer.[7] For the

(van Atten 2002a, 5), then there must be another route to mathematical objectivity, which could be *intersubjectivity* (Rahman 2016).

[6] If one considers that the meaning of a word in common language is determined through an infinitely proceeding sequence of actual decisions, as it is indicated namely in Stanley Cavell's reading of Wittgenstein's *Investigations*, then the problem of finding a satisfying constructive semantic for expression of free choice sequences appears as a crucial task. According to S. Cavell, «to say that a word or concept has a (stable) meaning is to say that new and the most various instances can be recognized as falling under or failing to fall under that concept» (Cavell 1979, 185).

[7] «For the hearer, in whose momentary field of vision the thing that we wish to point out is perhaps not present, only this indefinitely general thought is at first aroused: Something is being pointed

latter, it can be the case that only a general and indeterminate thought is awakened by a deictic expression like "this": there must be a complementary "presentation" for the determinate meaning to be truly and wholly constituted. Things go otherwise for the former: the indication itself is given as concretely determinate. Therefore, the hearer can gain direct access to the general compound of meaning, but not to its determinate and singular compound, thereby requiring a complementary representation, which must be provided indirectly by the speaker itself. For the speaker, *indicative* and *indicated* meanings are unified; for the hearer, they are not. Such a distinction between different ways of constituting the meaning of deictic expressions indicates that, when an ideal replacement through non-deictic expressions is not feasible, then one must acknowledge an intersubjective feature in the act of meaning constitution. This is the case at least from the perspective of the hearer, who needs to refer to the speaker itself in the determinate way he aims at the complete meaning of the fluctuant expressions.

If one wants to develop the content of this late *addendum* in relation to the posited irreducibly subjective compounds of meaning, one could say that this acknowledgment of asymmetry between the speaker and the hearer implies that both *interact in a play*, according to implicit rules, such that the hearer is allowed to ask the speaker for specifications in order to more completely understand the precise meaning of what the latter is expressing. Such a play would be a sort of an *alternative way* for a meaning-act: a *real* path to compare with the *ideal* way which is accessible only for the speaker, because he is the only one whose act of constitution is determined through a direct act of perception. The hearer knows that something is indicated, but he cannot know exactly what it is, and the way he aims at the indicated meaning supplies both a place and a *constitutive role* to the speaker itself in a *dynamic interaction*. The ideal *content* of both types of act (from the speaker and from the hearer) has to be exactly the same, but only the speaker's type of act provides a direct—and so to speak, an "authoritative"—access to it. In the case of the hearer, there must be a place, in the very act of meaning-constitution, for the speaker himself, through the way he provides access to the concrete meaning he is aiming at. The speaker is then an essential parameter in the constitution of the meaning of subjective expressions, which can hold as the product of what (Reinach 2012) will call, in 1913, a *social act*. In such an act, not only is attention paid to the one of the roles, but both are being addressed. In order to get the meaning of what the (proponent-)speaker is expressing, the (opponent-)hearer has to ask him for specifications. Through the answers provided by the former, the latter can constitute the targeted meaning, which *ideally* corresponds to the directly *indicated meaning* the (proponent-)speaker is occasionally expressing, without this ideal meaning being ever actually reachable. If the (opponent-)hearer cannot access the ideal meaning of deictic expressions and can only access a reconstruction of it through interactions with the (proponent-)speaker, it could be the case that a *real* process of meaning-constitution supplies the lack of access to its ideal determinate unity,

---

to. Only when a presentation is added (an intuitive presentation of the thing at stake demands an intuitive pointing out), is a definite reference constituted for him. For the speaker, there is no such sequence: he has no need of the indefinitely referential idea which function as «index» for the hearer» (Husserl 2001b, 199).

in such a way that it becomes necessary to conceive a kind of meaning-constitution which is immanent to the relation itself between the (proponent-)speaker and the (opponent-)hearer. Conceiving the meaning-constitution in the case of the *real use* of essentially occasional expressions in this way is obviously not *explicitly* stated in *Logical Investigations*. But one can claim that if, on the one hand, *ideal meaning* is inaccessible to the hearer in the case of deictic expressions, and if, on the other hand, there is actually a *real process* of intersubjective transfer implying social acts, then there must exist some kind of *real process of meaning-constitution* which cannot ever overpass *an essentially provisional way of being*.

In the case of free choice sequences—which Du Bois-Reymond did conceive for the first time in 1882, considering real numbers with decimals determined by throwing a die[8]—, this simply means that their potentially infinite development is conveyed by intersubjective transfer itself (which must then be potentially infinite itself[9]).

4.3 Occasional Judgments and Transcendental Intersubjectivity

Husserl would later say that he had been lacking, at the time of writing *Logical Investigations*, the means needed to conceive correctly the possible intersubjective constitution of meaning for the case of deictic expressions (Husserl 1969, 199). Nevertheless this wasn't a problem that disqualified him from developing the precise perspective of 1901, whose basic project was to study *what logic is about*. What he then assumed, as previously outlined, was an *idealism of meaning* regarding *real processes* as not determining meaning-constitution essentially.

If one considers the same point from the later perspective of *Formal and Transcendental Logic* (1929),—maybe under O. Becker's influence (Gethmann 2002, 112)—Husserl puts an anti-Platonist accent on his own conception of meaning through insisting on the idealizing acts of a (transcendental) subject, things deserve further investigation.. In a note, Husserl explicitly states that, in *Logical Investigations*, he lacked the "doctrine of horizon intentionality about which *Ideen*… have shown for the first time the determining role" (Husserl 1969, 199). It must be recognized that Husserl does not use "occasional *expressions*" anymore but rather "occasional

[8](Troelstra and van Dalen 1988, 21).

[9]An important point is that, if any complete substitution of objective expressions to subjective ones is unfeasible, then such a dynamical interaction can have no end in itself: the way the speaker constitutes the concrete indicated meaning through an act which is determinate through perception cannot be completely transferred to the hearer when he has no access to the same perception. It has an important consequence. To consider that intersubjective transfer has been satisfyingly accomplished could then depend on an *agreement* of both parts about a possible degree of determination of meaning. There would be a possibly infinite process of transfer through interaction which would forbid any complete determination of meaning through its expression, unless the norm of determination is claimed to be provided otherwise than through an (unreachable) "ideal unity" of meaning.

judgments"[10] which have "their own intersubjective truth and falsity"(*ibid.*). Further examination of the way the problem is considered in 1929 shows that the irreducible difference, that appeared in 1901, between the constitution of the meaning of a deictic expression for the speaker and for the hearer is not maintained as such, precisely because the concept of *horizon-intentionality* provides a means to neutralize their asymmetry.[11] Our proposal in the present paper is, so to speak, to return to the first attempt, which can be extracted from the *Logical Investigations*, by adding Reinach's view on social acts, as containing potential analyses which are better suited for the constitution of a non-idealist constructive semantics for the expression of *infinitely proceeding free choice sequences*. According to Husserl, when occasional judgments are expressed, the speaker and the hearer both rely on *similar types of situations*, which precede judgments themselves, and provide a fundamental level of evidence. It means that, for the speaker himself, there remains potentiality within the horizon of his own acts of perception. If the objective content of meaning is aimed at through a horizon-intentionality, the speaker and the hearer can both share the same type of constitution-act, so that the intersubjective transfer can be based on the similarity of their *experienced world*.

Husserl's treatment of intersubjectivity is now based on the claim that it is essentially possible to constitute any act of meaning-constitution performed by another subject through one's own transcendental subjectivity. Every potentiality for meaning-constitution is accessible to any subject. Therefore, if one uses an essentially subjective expression in order to express a judgment, the perceptual horizon out of which such a judgment is constituted as meaningful is similar enough to that of any other subject to provide the possibility, that anyone, looking at his own perceptual horizon, achieves the representations he needs, to complete the *indicative* meaning with an *indicated* one. The latter is then actually constituted as a *horizon* rather than as a determinate "*indicated meaning*" (but a horizon can hold as potentially determinate and as such possessing a fully determinate meaning).

Such an insight opens up the way to a treatment of intersubjectivity as something whose own constitution can be gained starting from a "*transcendental solipsism*" (Husserl 1969, 241) and arriving at "transcendental intersubjectivity".

Solipsism *on its own* can be claimed the "dead end" of intuitionism as a philosophy of mathematics. The problem stems from the obvious difficulty in simultaneously conceiving of mathematical objects as essentially resulting from *subjective acts* (i.e. from acts of a *Creating Subject*) and mathematical objects as being potentially trans-

[10]In the *Logical Investigations*, a judgment has been defined as an act of meaning associated with an act of fulfilment, possibly deceptive in the case of negative judgment. Husserl carried out accordingly a "doctrine of pure morphology of judgments", which has been further investigated in *Formal and Transcendental Logic*, notably through the uncovering of "*hidden intentional implications* included in judging and in the judgment itself as the product of judging" (Husserl 1969, 204). This uncovering approach is what led Husserl to the deep analyses provided in *Experience and Judgment*, pursuing in the world of life the very deepest ground of evidence.

[11]Maybe we can say that another direction could have been taken from such a distinction.

ferred from one subject to another without loss.[12] M. van Atten's decisive proposal to "identify intuitionism's Creating Subject [...] with Husserl's transcendental subject, as far as mathematics is concerned [...] promises [, according to him,] to solve the problem of how to account for intersubjectivity [...]" (van Atten 2002a, 80).

4.4 Difficulties

The question is how Husserl gives such an account of intersubjectivity, especially for the case of occasional judgments. Following the note mentioned above, this possibility of intersubjective communication is not based, in Husserl's view, on any intersubjective dynamical constitution of meaning, but rather on the basic communality of the *experienced world*, that comes before any determinate subjective differentiation, so that *subjectivity* has to be capable of being broadened towards *intersubjectivity* from the beginning. If meaning can be subjectively constituted, then it can be communicated, because the basis on which transcendental subjectivity constitutes itself necessarily has to be the same basis on which another subject will be constituted *as constituting himself.*

In such a framework, the most important principle of communication seems to be basic similarity of the very first potential horizon of the world of experience, which precedes any reflexive act of judgment. Interaction itself between different subjects does not have to be constitutive of the meaning of the expressions. There is, so to speak, a *common ground* of meaning which is immanent to the very first (common) world of experience, which is *lived through* before becoming a possible object of any predicative description. The way (Husserl 1969) sets up the question of intersubjectivity nevertheless goes hand in hand with the assumption that it is possible to carry out the *transcendental reduction* through a total abstraction of other subjects. In such a device, acts of self-constitution as a transcendental subject do not depend on the factual presence of other subjects, whose very existence has been *phenomenologically reduced.* If there are no remaining *concrete* relations to others to be taken into account, then the problem of the intersubjectivity of occasional judgments disappears, because occasional judgments derive their essential content from the ground of evidence which phenomenological reduction reveals and which is the very common ground of the world of life. Hence, following M. van Atten's

[12]This way of framing the problem presupposes that meaning comes before communication, which is Husserl's perspective. Conversely, it is possible that meaning could be constituted, from the beginning, through acts of communication, so that, for instance, the reference to a unique scale of time, according to which choice sequences are expressed without using deictic expressions, would be the result of a dynamic interaction. According to this perspective, the question is no longer: "how a meaning, which has been already intentionally constituted, can be, subsequently, intersubjectively communicated?" but rather "how a meaning can be, from the beginning, intersubjectively constituted?". When communication is viewed as the basis of any meaning, the question of the possibility of communication through deictic expressions replaces, so to speak, the question of the possibility for the meaning of such expressions to be determinate on its own. Such a shift has been notably accomplished within the so-called *inferentialist* semantic framework (Brandom 2000).

shift from Brouwer's own philosophical bases of choice sequences into Husserlian phenomenology: the creation of a free choice sequence is nothing other than constituting a temporal object according to determinate intentional structures, that anyone can identify.[13] The concrete being of a determinate sequence is nothing essential to this temporal object, though, according to its very essence, it is labelled with a determinate constructive agent, which cannot transmit what we could call his own constructive authority.[14] In other words, if another subject starts from the same initial segment, and then goes further on, *it is not the same choice sequence*.

On this precise point, there is something that remains unclear for us: on the one hand, a choice sequence is assigned to a determinate singular and irreplaceable subject, because another subject cannot possibly take its place to continue with the choices the other one has started with. On the other hand, every construction of a choice sequence is attributed to *the* Creating Subject and can be realized, including its open horizon, by any subject. A possible answer would be, as it has been suggested to us by M. van Atten, that, at least in the theory, the Creating Subject is thought of schematically, and in arguments one appeals to instantiations—just as the Universal Turing Machine has no history of computation, but is a schema that is instantiated in particular arguments. According to Brouwer, even though each mathematician has to make his or her own constructions, mathematics is the same for everyone. See for instance (Brouwer 1975, 99, 106).

This aspect certainly merits clarification, particularly in concerning the difference between an *actual* construction and a *potential* one. The question is how a determinate subject can hold as singular, despite the fact that it has to be constituted as such from the perspective of the transcendental subject itself. It is possible for the transcendental subject to constitute others as *possible forms* of being, transcending its own particular being, but it cannot constitute them as *actually being*. They can only be *potential* ways of being others. Such a point is clearly addressed by Husserl himself: Intermonadic structure only delivers a frame for *ideal possibilities*.[15] In such a frame there is no place for "contingent factuality".

---

[13](van Atten 2007) indicates the way intentional acts constitute such formal temporal objects in the chapter called «The constitution of choice sequences».

[14]"Choice sequences depend on the subject (who has to make the choices)" (van Atten 2002a, 34). Though (van Atten 2007, 96) mentions the "Kreisel-Troelstra translation, which shows that sentences quantifying over lawless sequences are equivalent to other that do not, and whose mathematical nature goes unquestioned", he also writes that "these [axiomatic] systems [of translation] should not be confused with the sequences themselves. Axiomatizations are a way to present mathematical content, but they are not identical with it" (41). See also (98): "[...] it is essential to a lawlike sequence to have a bearer or an owner, and such a sequence cannot be hand over from a subject to another".

[15]Husserl (1995), Ricoeur (2004, 271). See also Bassler (2006, 595): "the strategy for defending intuitionistic mathematics as intersubjective is predicated on establishing the *identity* of the subject with respect to the structure of the transcendental ego" and (Kern 1964, 205): "Auf dem Cartesianischen Weg wird die fremde Subjektivität nur als "blosses Phänomen" erreicht".

4.5 Eidos and Ousia

In order to avoid this difficulty arising from the essentially "eidetic" way of treating intersubjectivity, which is such that the very facticity of choices cannot be given a determinate status, it is certainly relevant to follow O. Becker in the deepening of his own approach by *free choice sequences*, in *Mathematische Existenz*, moving away from Husserl's straightforward approach to phenomenology.

The necessary result of the "eidetic reduction" is the bracketing of facticity. Yet, in Becker's view, another way exists in phenomenology, that does not rule out facticity. In a late article called "Zwei phänomenologische Betrachtungen zum Realismusproblem", he starts redeploying the multiple possible paths Husserl opened from the outset through the idea of phenomenological reduction. According to Becker, the full reduction decomposes itself into two independent compounds: the *transcendental* and the *eidetic* reductions (Becker 1962, 12). In the framework of eidetic phenomenology, there is hardly any place for facticity. But that is not the case in pure transcendental phenomenology, which suspends any knowledge of the world, while maintaining facticity and contingency within phenomena. In a reductive method such as this, an individual's history is *seen*, but not *conceived as an eidos*. This means that there is a kind of reduction which conserves singularity of events, as long as eidetic reduction does not guide the transcendental investigation. According to Becker, Husserl has never followed this path, but he has occasionally indicated its existence and the associated tasks it involved. By contrast, Heidegger (and Scheler) have, according to Husserl himself, gone down such a path (Husserl 1995, 160).

The approach to choice sequences, which is deployed in the sixth chapter of *Mathematische Existenz,* corresponds clearly to such a non-eidetic transcendental reduction,[16] and will perhaps be helpful in clarifying what is at stake here. Becker claims that the decisive feature of intuitionism in the dispute against Hilbert's formalism, is that it contains the essential possibility to investigate the existence of mathematical objects, whereas the formalists' definition of existence through consistency contains a rejection of it. Intuitionism is based, according to Becker, on the postulate that every mathematical object has to be possibly attainable through factual syntheses. Accordingly, Becker seeks an anthropological foundation ("*Fundierung*") of mathematics, which consists in the factual life of humans ("*das faktische Leben des Menschen, das jeweils eigene Leben des Einzelnen (oder wenigstens der jeweiligen* "*Generation*") *ist das ontische Fundament, auch für das Mathematische*") (Becker 1927, 636).

Free choice sequences are considered by Becker as being anchored in the factual individual existences, which project themselves forward into the full obscurity of

---

[16]Becker mentions the distinction, provided in ancient Greek, between εἶδος and οὐσία, and situates his own path to a hermeneutical study of the facticity of mathematical construction as dealing with the latter (to compare with Husserl's eidetic way). There can be no eidetic analysis of facticity, though there can be "characteristic features" (*charakteristische Züge*) of facticity". In this case, "es handelt sich um eine besondere Art von Begriffen, nämlich um formal "anzeigende" (Heidegger), deren "Allgemeheit" in ihrer Bezogenheit auf das "Jeweilige" liegt." (Becker 1927, 627).

the future. One can actually wonder if there can still be a way to conceive the communicability of such purely temporal objects, which in their very singular way of being—in their *ousia*—do not comply to the rules of any general formal scheme, in the sense of laws but maybe also in the sense of a *type*.

If intersubjectivity can be yet conceived of, for the expression of *free choice sequences*, despite such a radical view, based on their being *contextually occasional*, then we will gain new insights into the question of a semantics for the Creating Subject.

4.6 Back to Descartes: Another Way to Intersubjectivity

We will see below how, in a small paper called "Husserl und Descartes", (Becker 1936) sketches in a few words an interesting *alternative to monadological intersubjectivity*. Nevertheless, further elaboration on the monadological way will be necessary beforehand, in order to understand the very nature of the shift. The main feature of the way traced in the fifth *Cartesian Meditation* is, as previously outlined, to hold as a frame for ideal possibilities structuring any relation to others (so that such a structure rebounds on the necessity-relations of objective nature, as nature holding for any possible human being). In the monadological intersubjectivity framework, other men are nothing but *potential*, so that any insight into structures essentially regulating human interactions is based on the ability to *imagine* potential ways of being, which are attributable to others. *Actual* ways of others' being are accordingly nothing but particular instantiations of *potential* (or *ideal*) ones. In a bracketed natural world, intentional acts *potentially* constituting objects are the same as intentional acts *actually* constituting objects: there is no "rest" beyond intentionally constitutive acts of reduced phenomena.

There is naturally an essential difference between the constitution of another human as *potential* and the constitution of it as *actual*, but the potential constitution of a human is the same as the actual constitution of it (*constitution of something as potential* is not the same as *potential constitution of something*). Now it is important to note that while Becker would in a sense disagree with van Atten's claim that in Brouwer's intuitionism, as in Husserl's transcendental phenomenology, transcendental possibility is equivalent to existence with respect to purely mathematical objects (van Atten 2007, 95), he at least would say that such a restriction of choice sequences to their purely *mathematical* dimension neglects what constitutes their factual dimension. Or better: Becker adds the perspective of a hermeneutic ontology here. According to him, mathematical objects do not only have to be transcendentally possible, but also have to be effectively constructed in order to be potential objects of judgment.

In an intermonadic transcendental framework, mathematical creative constructions essentially consist of the exposure of transcendental possibilities, though transcendental possibilities can only be known by carrying out the construction. Communicability accordingly is a transcendental possibility of "intermonadic community".

Following M. van Atten's reading, a free choice sequence created by the Creating Subject would therefore essentially be a constructive possibility whose particular feature is to be temporally determinate, possibly in accordance with a unique scale of time[17] any concrete subject can rely on. The accorded interpretation of intersubjectivity for mathematical constructions can definitely be brought together with Husserl's intermonadological conception. But such a restriction to constructability implies a full fade-out of the factual dynamical dimension of subjective acts that Becker considers, by contrast, as a central feature of the intuitionist approach. A monadological intersubjectivity-framework deals with *possible acts of constitution*, which then presuppose that fully determinate meaning has to precede any act of meaning-constitution. Accordingly, there can be no creation of meaning and no springing out of new meanings through original acts/decisions of a subject, because meaning is thought of as being omni-temporal. Strong revisionism is accordingly not really a creation of original mathematical possibilities, but only an original acknowledgment, through transcendental investigations, of possibilities of constructions, eventually contradicting the *factual* mathematical practices.

In order to conceive a true creation from nothing (which maybe implies a movement from a reading of Brouwerian intuitionism as it has been furnished, for instance, by Gödel[18]), that is a creation which generates its own transcendental possibility, a *shift* from Husserl's eidetical framework is necessary. Only then, there will be a place for creative facts.

It is precisely such a shift which is sketched in Becker's "Husserl und Descartes", through an original way of accentuating the imaginative function in the eidetic variation device.[19] Becker suggests that there can be a way of being, springing out from

[17]According to (van Atten 2007, 90) the meaning of propositions about choice sequences is *not* intra-temporal whereas choices sequences are. "Consider, for example, the concept expressed by 'The lawless sequence begun by me exactly one year from now at 6.58 pm'. The meaning of this expression, and hence the concept, is invariant over all time, past, now, or future. It does not change; it is omnitemporal " (89). It is actually very close to Dummett's proposal to "regard time, from some fixed point on, as being divided denumerably into discrete stages" in order to define a "sentential operator" corresponding to the *Creating Subject.* Our proposal is to consider that such a "naturalized time" obliterates the dimension of meaning which is irreducibly based on the intra-temporal anchoring of factual constructions.

[18]"The central and appropriate concept for Brouwer is construction rather that creation" reported in (van Atten 2002a, 9).

[19]Husserl claims that «to seize upon an essence itself, and to seize upon it originarily, we can start from corresponding experiencing intuition, but equally well from intuitions which are non-experiencing, which do not seize upon factual existence but instead are «merley imaginative» (Husserl 1983, 11). Accordingly, «positing of [...] essences implies not the slightest positing of any individual factual existence; pure eidetic truths contain not the slightest assertion about matter of facts» . If one bypasses the eidetic reduction, accomplishing only the transcendental one, there can actually be a place for factual limitations of the depth of imagination. This then permits discussion of the factual existence of others through pure encounters which broaden this depth. Therefore, eidetic reduction could depend on the implicit positing (inside a transcendentally reduced world) of individual factual existences, from which the very depth of imagination itself would be constituted. If eidetic intuition comes from the free use of imagination, it can be concluded that it is based on factual encounter events, which bound from the beginning the imagination's depth to the very ground

others, which I could absolutely not have anticipated, "*weil ich dazu nicht reich genug bin*",[20] that is because the strength of my own imagination could not go far enough to constitute by itself the very originality of other humans. Here, the problem is not a lack of a sufficiently high degree of imaginativeness, but rather a lack of imaginativeness of the right kind: there can be no act of my imagination that constitutes the originality of another human being. It can be concluded that the posited dependency of eidetic reduction on imaginative power overestimates, in Becker's view, such a power. Even if the way another human being constitutes an object through constitutive acts is understandable to me, it doesn't necessarily imply that such a method of constitution *actually* has its fully determinate counterpart in my own subjectivity, in the sense of there being a pre-established harmony between our respective intentional acts. It is possible that my own actual execution of transcendental acts essentially depends on the real being in front of me, of such a human, whose behavior would, for instance, at first appear as an enigma to me, so that I would have to *neutralize* my own way of constituting objects in general, in order to make possible any (argumentative) interaction with him.[21] Hence, the depth of my own imagination would depend on

---

of intersubjective events. Eidetic intuition would then be a matter of reconstituting the domain of possible things and possible thoughts through a schematic remembering of factual events that have constituted and broadened, step by step, «subjective faculties» themselves. Then imagination could be intersubjectively constituted in the sense that its depth cannot be conceived of as being spontaneously increasing, but rather as being constructed from pieces of impactful encounters in one's life. The imaginative power of an individual mind would be based accordingly on real processes and events in individual history: it would be the result of a secondary internalization process.

[20](Becker 1936, 621). The basic idea is to follow a cartesian path (that is neither Leibnizian, nor Kantian) inside the phenomenology of intersubjectivity, based on Descartes's third *Méditation*, especially on the distinction between "réalité formelle" and "réalité objective" (Descartes 1999, 437–438). This path is investigated more precisely in (Becker 1962, 16–17). According to (Brandom 2000, 46–47), one of the most important improvements that Leibniz, among other post-Cartesians, furnished within "enlightenment epistemology" is the claim that the content of representations must be constituted through the inferential role they play in a "space of reasons" (Sellars), which must then come antecedently. It could be asked to what extent Becker's view on intersubjectivity depends on Descartes' *representationalism*. In order to avoid such a presupposition, the facticity of the other's presence could be conceived as the *permission*, within a play, to express spontaneous requests, whose possibility would exceed any predictability through explicit rules, without being in contradiction with it. Openness to new meanings through the *representation* of spontaneous acts, coming from others, would also be a feature of the *frame* within which intersubjective interactions would be regulated and in agreement with which some inferences could be considered as being based on *choices*.

[21]Actually, further investigation needs to be carried out. It is possible, that the exploration of the potentially infinite horizon of one's experienced world requires, from the outset, dynamic interaction with another one. To begin with, if experienced world holds as the very last product of *phenomenological reduction*, it could be claimed that access to it cannot be gained without intersubjectivity. In *Zwei phänomenologische Betrachtungen zum Realismusproblem*, O. Becker suggests that one of the purest form of phenomenological reduction can be indicated through the situation of "dialogue partners in a philosophic or scientific discussion":

"In einer solchen müssen die sich Unterredenden von ihren Überzeugungen, von dem von ihnen eingenommenen Standpunkt abstrahieren, ihn für den Augenblick (ohne ihn innerlich preiszugeben) in der Schwebe lassen, um überhaupt argumentieren zu können. Man darf ja, will man einen wis-

the very facticity of the concrete *meeting* of such a human. The particularities of the other's behavior would mark the occasion without which my own imagination could not have gone in any determinate directions.

It follows that there is a place for deictic expressions at the very basis of mathematics itself, if mathematically acting subjects in front of me are carrying out operations which I would absolutely not have thought of by myself and which constitute the fundamental basis of my own mathematical activity and development. If the mathematical activity of creation has thus to be essentially situated, then occasional expressions make up the basic vocabulary according to which creation as such has to be expressed. The domain, from which a term would derive its meaning, would be shaped through creating gestures, springing out from factually new practices and singular manners of acting.

4.7 Interlude: Brouwer on Intersubjectivity

In Brouwer's own discussion of mathematical creation, there actually is such a use of deictic expression, which arises notably in a 1946 text in praise of Gerrit Manourry. That might hold as characteristic of a concrete intuitionistic way to see what comes to the fore in mathematical concrete practice. It takes the form of a personal account:

> As happens so often, I began my academic studies as it were, with a leap in the dark. After two or three years, however full of admiration for my teachers, I still could see the figure of the mathematician only as a servant of natural science or as a collector of truths: -

---

senschaftlichen Beweis führen, nicht das zu Beweisende schon voraussetzen, ehe man seinen Beweis zu Ende geführt hat. Ein solches In-der-Schwebe-Lassen (die sogennante Epoché) [ist] nun gerade die Haltung, die von der phänomenologischen Reduktion gefordert [ist]" (Becker 1962, 3).

It is then as if it was necessary to assume the perspective of someone else, basically disagreeing with what we intend to prove, in order to get the determinate type of relation to things which is required by phenomenological reduction and which presupposes a (provisory) neutralization of all determinate belief about concerned things. It can be claimed that the (provisorily) radicalest degree of reduction depends on the conceivable disagreement of others with the basic convictions we have, and that radicality of reduction is, from the beginning, constituted through intersubjective interaction. If, on the one hand, meaning has to be rooted in "experienced world" or "life-world", which provides normative insights and original experience that come before any act of judgment, and if, on the other hand, access to this world as such is provided through *phenomenological reduction*, then meaning elucidation of any expression (and not only the essentially occasional ones) depends, in Husserl's view, on the device through which *reduction* itself is carried out. It can furthermore be claimed that there cannot be any *definitive fundamental level* beyond which phenomenological reduction could not go deeper, in relation with possible intersubjective disagreement on the most fundamental concerns. As an obvious consequence of such a claim, it must be the potential (opponent-)hearer who determines the attended degree of determination of meaning, given that there can be no conceivable such degree for meaning *in itself*. Any claim to reach the very deepest level of phenomenological evidence (and so the very first act of objective meaning constitution) is threatened by an irreducible openness to possible further steps into radicality. In Becker's view, Heidegger's so-called phenomenological Destruction aimed, through historical-critical observation, at destroying the ineradicable prejudices ("die unausrottbaren Vorurteile") (Becker 1962, 4) that phenomenological reduction would always fail to overcome by itself.

> truths fascinating by their immovability, but horrifying by their lifelessness like stones from barren mountains of disconsolate infinity. And as far as I could see there was room in the mathematical field for talent and devotion, but not for vocation and inspiration. Filled with impatient desire for insight into the essence of the branch of work of my choice, and wanting to decide whether to stay or go, I began to attend the meetings of the Amsterdam Mathematical Society. There I saw a man apparently not much older than myself, who after lectures of the most diverse character debated with unselfconscious mastery and well-nigh playful repartee, sometimes elucidating the subject concerned in such a special way of his own that straight away I was captivated. I had the sensation that, for his mathematical thinking, this man had access to sources still concealed to me or had a deeper consciousness of the significance of mathematical thought than the majority of mathematicians. [His papers] had the same easy and sparkling style which was characteristic of his speech and, when I had succeeded, not without difficulty, in understanding them, an unknown mood of joyful satisfaction possessed me, gradually passing into the realization that mathematics had acquired a new character to me. For the undertone of Mannoury's argument had not whispered: "Behold, some new acquisitions for our museum of immovable truths", but something like this: "Look what I have built for you out of the structural elements of our thinking. – These are the harmonies I desire to realize. Surely they merit that desire? – This is the scheme of construction which guided me. – Behold the harmonies, neither desired nor surmised, which after the completion surprised and delighted me. – Behold the visions which the completed edifice suggests to us, whose realization may perhaps be attained by you or me one day (Brouwer 1975, 474–475).

Though this description obviously presupposes a certain type of intersubjectivity for mathematical construction, one could object that such a text, a kind of panegyric or personal tribute, should not be taken to be of central philosophical importance for Brouwer's thought, especially if one considers that in Brouwer's view no mental construction can be perfectly rendered through language so that an inaccessible rest remains "inside" the constructing mathematician.

4.8 Towards a Dialogical Reading of the Creating Subject (and Another Interlude)

In contrast to the transcendental turn of phenomenology, we claim that the differentiation of roles that Husserl first outlined in *Logical Investigations* along the lines of Reinach's view on "social acts",[22] could provide a more precise understanding of Brouwer's own view on intersubjectivity. To reiterate: Intersubjective interaction should not be thought of as intermonadic harmony of transcendental intersubjectivity (which might have to be established in a dialogue, with another subject or with oneself), but as a dynamic interaction through which the hearer asks indefinitely for specifications from the speaker who has been using a deictic expression, so that the latter has the duty to provide the precisions which are requested by the former. That means that the incomplete perceptible object that the mathematician points to, for example on a blackboard, saying "behold what I have constructed", would not be communicable through the similarity of the "experienced world", but rather through

[22] Which is certainly to complete through a scrupulous reading of (Husserl 1973b), and most prominently of the texts called "Gemeingeist I" und "Gemeingeist II".

the act of providing, through speech acts, the precisions which are requested by the hearer to whom the communication has been addressed.

It is important to mention that it does not really matter *who* is providing such precisions: it can actually be the hearer projecting himself into the speaker's role. The point is that openness to new constructions has to come from elsewhere, so that a new space of possible reasons emerges. *It has to* because dynamic construction of meaning through speech acts has to be a *reaction* to someone's original way of talking and acting, that is a reaction to a autonomous dynamics that no spontaneous use of my imagination could ever constitute.

We could say that a mathematician's claim to have constructed something new is a kind of *promise*,[23] which implies a certain distribution of duties and entitlements concerning speech acts, starting from an initial claim. Even if only the (proponent-) speaker is supposed to have "direct access"[24] to his "interiority", to the meaning that he is expressing (and to a corresponding object he occasionally points to), the external accessibility of this meaning can be warranted through speech interactions between him and the (opponent-)hearer, in such a way that, precisely through this interaction, the former "neutralizes" his own support of the meaning and recovers it, so to speak, from the point of view of the latter. Therefore, the ideal unity of meaning which is supposed to guide the construction (and, in the case of a free choice sequence, which modifies itself over time) can be thought of as being nothing else than the potentially infinite *tree* (or *spread*, to use Brouwer's term) of all possible interactions based on sequences of speech-acts' requests addressed to a (proponent-)speaker.

To use concepts from the framework of dialogical logic, which has been developed as a continuation of the work done by the *Erlangen school*, in order to offer a constructivist approach to meaning, and which is now prominently deployed by S. Rahman and his collaborators, we can say that at this level, called the *play level*,[25] the question of *objectivity* is not relevant: meaningful expressions do not necessarily have to be fulfilled or even be fulfillable by a subjective constitution. The construction of objects *fulfilling* signification-intentions has to be situated on another level, which can be called *strategic level*, through the *possibility* of a *winning strategy*, that is a way to defend a *posit* (a proposition assumed to be true) against every possible challenge coming from possible (opponent-)hearers.[26] The *dependency* of such pos-

[23]The precise difference between a *posit within an intersubjective space of reason* and a *promise* would still deserve scrupulous investigation.

[24]Provided that it has sense to talk of such a direct access, which certainly requires a criterion to ensure its existence.

[25]In the dialogical framework, deployed by S. Rahman, what "characterizes the play level (of non-material dialogues) are speech-acts of acceptance that lead to games where the proponent, when he wins a play, he might do so because he *accepts* some specific moves brought forward by the opponent during the play (without asking for the evidence that supports the proposition brought forward)—this leads to some kind of pragmatic-truth, if we wish to speak of truth" (Rahman et al. 2016) A systematic study going through the deepest issues of the distinction between play level and strategic level is to be found in (Rahman et al. 2018).

[26]"The strategy level is a level where the proponent wins accepting whatever the player might posit in every play that constitutes that strategy".

sibilities on the *factual* springing out of determinate "challenge positions" in one's world of experience relativizes the *existence* of this strategy, and also *intersubjectively relativizes* the existence of the constructions the mathematician is pointing to. But such an intersubjective relativity does not imply *relativism*, because if a winning-strategy does exist *in possibility*, it means that there is *absolutely* no thinkable way to deny the corresponding construction. The springing out of a *decisive* refutation would hold as a proof of the lack of a winning-strategy. The point is that the *norm* of constructability of propositions does not have to be some kind of *ready-made* transcendental subjectivity: it can be the dynamically open intersubjectivity itself, as long as it produces its own immanent rules.

The radicality of phenomenological reduction, which provides the deepest roots of any "meaning explanation" beyond linguistic explanation essentially depends on the radicality of concrete speech-acts addressing different modes of request. *Eidetic variation* is thus essentially rooted in what we could call *intersubjective variation.*[27]

Such a view opens up a new way to conceive the very essence of the *Creating Subject,* that can be thought of as an intersubjective principle of construction developing itself through effective social interactions to which every human act contributes, as long as it delivers possible means to satisfy what we called "precision requests" or even means to originally formulate such requests. Deictic expressions are, so to speak, the basic way of pointing out any theoretical concerns. The basis of creation then is the very act of saying, by pointing to something perceptible, indicating that "there is something *more* there", exactly in the same way a free choice sequence always potentially delivers more than what has been constructed. The latter serves as a paradigm of any free creation of sense. In this view full determination of meaning for indeterminate objects like choice sequences hasn't been provided for by any ideal unity *in itself,* but by implicit rules regulating the speech interactions, opening an internal space, whose limits do not contradict its infinite openness to unforeseeable speech-acts.

In the extract quoted above, Brouwer mentions that the whispering "*undertone*" of Mannoury's argument is like an invitation to look at something which has been constructed out of structural elements of ("*our*") thinking. It certainly can be brought together with the manner in which Brouwer talks about the possibility of mutual understanding in "Consciousness, philosophy, and Mathematics":

---

[27] Though it is impossible to explore, in any meaningful way, the subtleties of Kuno Lorenz's thought in a mere footnote, it can be mentioned here, in anticipation to further studies, that most prominently (but not exclusively) in the article called "Das Vorgefundene und das Hervorgebrachte", a way to a non-husserlian phenomenological reduction, which not only brackets the validity of the external world, but which also suspends all classification accomplishments (*Gliederungleistungen*) of consciousness, which appear in the real situations of learning, has been studied in great details. Moreover this phenomenological reduction is considered as overlapping with dialogical construction (Lorenz 2009, 171–172). It certainly deserves a specific study in relation to our present concern. Allow us to mention here, incidentally, the important Ph-D work of Mohammad Shafiei, which aims at rendering a dialogical account of meaning *within* transcendental phenomenology. M. Shafiei starts from a non-social account of the essence of dialogue, in order to avoid that it be "suspended by means of the transcendental *epoche*" (Shafiei 2018, 5). Whilst we share many of the same concerns, we are pursuing a different line of inquiry.

> Only through the sensation of the other's soul sometimes a deeper approach is experienced. And when wisdom[28] revealed by the beauty of this sensation finds expression in the antiphony of words exchanged, then there may be mutual understanding (Brouwer 1949, 1240).

Brouwer defines the "soul" of a human being as "the whole of egoic sensation indissolubly connected with [the corresponding] individual" (Brouwer 1975, 480). We assume that in Brouwer's view the real mutual *mathematical* understanding does not necessarily have to be thought of as a "transfer", from one subject to another, of an exactly determinate content, which has been fully constructed by the former, but rather as the opening of a *common* space of play, taking its departure from the very singular way a real individual is presently talking.[29] Meeting Mannoury is expressed in terms of the "well-nigh playful repartee" of his speeches and of the "sparkling style" of his papers, and in a broader sense in terms of "*the sensation that, for his mathematical thinking, this man* [...] *had a deeper consciousness of the significance of mathematical thought than the majority of mathematicians*".

Following Becker's suggestion, we can claim that the phenomenological meaning of the reality of others is provided through their unforeseeable singularity, which no

[28] It could certainly be said that the "wisdom" Brouwer associates with the "mutual understanding" is part of his mystical concerns, which have nothing to do with the communication of mathematical results. M. van Atten is certainly correct in claiming that "mystical insights are of no use in mathematics" (van Atten 2002a, 9). and that in the domain of wisdom, there are no mathematical constructions, because such constructions imply time and time keeps consciousness away from its "deepest home". Nevertheless, this does not contradict Brouwer's obvious attempt to evaluate mathematical activity from the perspective of wisdom (which is defined as "abolishment of discernment between the subject and something different" (van Atten 2007, 78). Let's consider the only "philosophical" text mentioned in "Consciousness, Philosophy and Mathematics": the *Bhagavad-Gîtâ*,. This is actually a poem which takes as its point of departure the situation of a warrior, Arjuna, to whom Krishna teaches that the way to wisdom can be the way of *action in the world*, because lawful occupations do not diverge from Brahman, but can be made a means of closest union, so long as we practice indifference to the fruits of action. A man can also act and nevertheless be "beyond action" (Prabhavananda and Isherwood 1956, 52). According to Brouwer, mathematics consists of nothing but acts of construction. Could it not be the case that Brouwer looks for wisdom through constructive mathematics? It is worth evaluating to what extent the way Brouwer situates mathematics in the context of eastern philosophy can be brought together with the teaching of the *Bhagavad-Gîtâ* concerning the way of Yoga through the acts. It is not absurd to say that there can be a practice of mathematics in accordance with wisdom if one considers that Brouwer writes, according to what he calls the second and third phases of consciousness (that is the *causal* and *social* phases), of the "activity of constructing things" which has to do with beauty when it is "exerted playfully", beauty being here obviously a sign of a way to "wisdom". It would certainly be fruitful to compare the chapter called "the Yoga of Knowledge" with the following thought: "The fullest constructional beauty is the *introspective beauty of mathematics,* where instead of elements of playful causal acting, the basic intuition of mathematics is left to free unfolding. This unfolding is not bound to the exterior world, and thereby to finiteness and responsibility; consequently, its introspective harmonies can attain any degree of richness and clearness." A parallel between Brouwer's thought and Indian Philosophy has been drafted by Roddam Narasina (Narasina 2009).

[29] It is actually the way G. Sundholm talks, indeed ironically, of Brouwer's view: "To everyone his own mathematical activity, and if, by transmitting beautiful constructions to other subjects, he can give someone pleasure, so much the better" (Sundholm 1984). The relation between the notions of "play" and of "pleasure" certainly deserves further investigation, particularly in regard to the perspectives deployed in the third Kantian *Critique*. A study of the relation between pleasure and

spontaneous use of imagination could ever predict correctly,[30] and which opens a future. Such a singularity cannot be expressed in another way than through the deictic indication of the way such individuals act. Accordingly, the creative experience would depend essentially on the way it finds its basis in the real-life relation to others.

We claim that the "solipsism" attributed to Brouwer emanates from the real meeting of singular personalities whose originality holds as the very basis of his *playful* constructions, that is constructions which are *not targeting any result*. In a broader sense, solipsism is nothing else than the (provisory) framework in which internal constructions can be carried out according to basic (internalized) requests expressed by singular individuals, whose practical accomplishments provide a measure of the level of "specification" which is required in order to produce effective constructions and whose muted "whispering undertone" outline the map of any meaning and claim to objectivity.

4.9 Paths to a Formal Approach

The *Creating Subject*, as an idealized agent of any mathematical construction is deictically anchored in the concrete constructing activities of humans, which request, by means of their "sparkling" way of being, a bracketing of the *ego*'s most elementary assumptions, only for them to be reconstructed from an unfamiliar *playful* perspective. It is now clear that the philosophical basis of some of the most recent developments in constructivism, especially Per Martin-Löf's *Constructive Type Theory*, in which the Creating Subject is not given a central place,[31] however consists of the conception of cognitive acts from a "first person perspective" associated with the claim that "every object come as type" (Van der Schaar 2011). One can wonder if the way P. Martin-Löf models expressions of free choice sequences in the framework he is using allows a deepening of the cognitive acts, which are phenomenologically primitive, by taking into account the intersubjective dimension[32] we are presently

knowledge has been provided in the frame of contemporary hermeneutical phenomenology around the *Critique of Judgment*, notably in "Système du plaisir (kantien)" (Nancy 2000, 65–80).

[30]"[...] actual human beings can instantiate, as far as it goes, the Creating Subject. But if, while doing so, they impose (further) limits on their (already limited) ability to create, they are no longer instantiating, however imperfectly, the Creating Subject." (van Atten 2015, 13). Accordingly, a reading of the Creating Subject as being intersubjectively instantiated would furnish the means to surpass the limitations of an individual's ability to create, through the expressions of individual request acts, within an open space of play, whose rules remain open to further specifications through *effective* new instantiations.

[31](van Atten 2018, 53–54) states that there is no place for the Creating Subject in Martin-Löf's theoretical approach.

[32]This is actually a line P. Martin-Löf followed in a recent lecture held in Oslo. The notion of «request» is here given a central place at «the very bottom» of logic, here called «the deontic level of logic» . The conclusion of the lecture is that «at its very root, logic is based on something which belongs to the area of ethics» (Martin-Löf 2017). We would like here to suggest that for a retrospective reading of the conception of choice sequences, deployed in (Martin-Löf 1990), it

sketching. Also Becker's proposal to distinguish between *transcendental* and *eidetic* reduction might be fruitful in this respect.

Following Troelstra (Martin-Löf 1990, 156) considers a choice sequence as a potentially infinite list of hypotheticals[33] that he expresses as follows:

$$\alpha = f_0(f_1(f_2(\ldots)))$$

It corresponds to an indefinite "non-well-founded" process:

$$\text{“}\alpha = \alpha_0 = f_0(\alpha_1),\quad \alpha_1 = f_1(\alpha_2),\quad \alpha_2 = f_2(\alpha_3),\quad \ldots$$

where f_i is a function from A_{i+1} to A_i, where A_i is a non-empty set, that is, a set containing an element a_i, for i = 0, 1, 2, etc."

At the beginning of the process, "we only know that α_0 belongs to the set A_0. Then, we ask what element of A_0 that α_0 is, we get to know that $\alpha_0 = f_0(\alpha_1)$, whereas α_1 belongs to A_1, whereas α_0 has become partially determined: it is neither freely variable nor constant but something midway in between." (156) The meaning of α is determinate through an essentially open parameter, which is symbolized through the dots "…". If we assume that choice sequences are essentially constructions from a first-person perspective and if the constructive act comes *factually* from a "precision request" enacted by an (opponent-)hearer in response to *my* initial indication of something corresponding to α, then the process of determination of my initial indeterminate expression corresponds to the impulse provided by such requests, rather than to my spontaneous indefinite examination of an indeterminate horizon.[34] Each

would be worth, starting from this recent deepening of logical horizon to the level of ethics. A systematic working out of the possible general links between CTT and dialogical logic can be found in (Rahman et al. 2018).

[33]But, as such, choice sequences are non-standard objects, that in the end are always eliminated from proofs, as suggested by (van Atten 2018, 53). A process's ending contradicts its potential infinity. The correlative positioning of an end, stated for an indefinite nonwellfounded process might rely on the brouwerian concept of a bar, and maybe to the dialogical concept of a rank.

[34]The idea that the determinateness of a potential infinitely proceeding explanation of meaning can be retraced to the special way Wittgenstein solves the problem of the infinity of decimals in real number expressions in his *Philosophische Bemerkungen* (p. 190). According to him, "Reelle Zahl ist das, was mit den Rationalzahlen vergleichbar ist" and therefore "Die eigentliche Entwicklung der Zahl ist die, die den unmittelbaren Vergleich mit den Rationalzahlen erlaubt" (Wittgenstein 1989, 236). In other words, though the decimal expression of an irrational number could proceed infinitely, it does not imply that this expression cannot be fully determinate, because one can find a principle of determination in the very position of a rational number, which the irrational is to be compared with—let us mention incidentally that Brouwer constructed in 1921 a decimal number which has (for the time being) no decimal expansion. Then, it is natural to conceive such a position as a request expressed, in the frame of an interaction, by an opponent, through the choice he makes of such an element of comparison. The infinite horizon, expressed through the dots "…", does not correspond

step into a further determination of α is the outcome of a request and takes the form of a choice whose concrete possibility is facilitated by the request itself. There must be a reason why the process is extended until a determinate step in order to make a full determination of its meaning possible despite its essential openness. If one thinks of such a process of determination in another way than as a pure *abstract* possibility (for example as a concrete process of word-definition) then it has to correspond to concrete requests. We claim that the way of being of these concrete requests is the way of being of the Creating Subject itself, as being based on intersubjectivity. Accordingly, every claim which is reconstructed according to an ultimate level of request, whose last possible degree of determination corresponds to pure variable can be thought of as being phenomenologically reduced (and determinable as fully meaningful).

The internal space of phenomenological reduction has to be delimited by concrete oppositions coming from real dialogue partners. It does not matter which kind of object is said to be non-determinable beyond such a limit. If someone claims to have access to a mental construction whose meaning cannot be shared with someone else, it necessarily means that he refers to a level of request that has been addressed to him by someone, having certain concrete practices (most notably linguistic ones), opening a normative space in which anybody can enter, provided such concrete practices can be really learned. Hence, there must be a way to indicate occasionally such practices by pointing to concrete human practices.

In our present view, free choice sequences could constitute the implicit shape of any proof-object[35] under the condition that eidetic reduction is bypassed, for no cognitive act can hold as a primary phenomenological fact. By contrast, free choice sequences would constitute a formal expression of a transcendentally reduced world, in which evidence is not "phenomenologically primitive", but is rather the very result of an intersubjectively provisory act of agreement about the degree of necessary improvement of meaning explanations.

---

to any actual infinite set, nor to a subjective power to explore this horizon still further, but rather to the concrete ability of someone to play the role of an opponent, by expressing a concrete request, which obliges any individual claiming their own "mental" possession of a corresponding potentially infinite construction to develop it until the level which makes possible the requested comparison. Though Wittgenstein refuses to consider *free choice sequences* as genuine mathematical objects (even though according to (Marion 1998) he has no convincing argument for it), his line of argumentation could be broadened to their potentially infinite constructions. Bernadette Dango, has defined, in the frame of her thesis concerning belief-revision a belief operator whose meaning is dialogically defined through rules, allowing an opponent to extend the context from which a proponent's belief is evaluated (Dango 2016, 270). In such a frame, free choice sequences could hold as purely formal expressions of defenses provided by a speaker defending the effectivity of the construction he claims to possess, through successive choices, which corresponds to explanation speech-acts. Developments of such meaning explanation sequences would be then regulated through the very existence of means to express opposing requests. Then creativity would be a property of intersubjectivity itself, because it comes as much from the expression of requests as from the answers given to them.

[35]Indeed, the insight that the Creating Subject's mathematical activity has itself a mathematically explicable structure is essential to the theory of the Creating Subject.

4.10 Construction and Being-in-the-World

In the general proposal we are presently outlining, expressions of free choice sequences must indeed be *embodied* in concrete intersubjective relations in order to get their determinate meaning. Deictic or occasional expressions of the real-life relational frame in which constructions are embodied accordingly constitute the deepest level on which any meaning can be determinate. Accordingly, ultimate determinations of meaning are anchored in something which cannot be *said*, but can only be *shown*, and which concerns the very facticity of the human being. Nevertheless, it does not mean that such an "anthropological grounding" of the meaning-practice obliterates its possible objective referentiality. The general idea is rather that the constructability of objects (even in the case of mathematics) gains its shape through concrete acts that spring out of singular gestures, accomplished by concrete humans, without any guiding abstract operative scheme, but in such a way that schemes constitute themselves pragmatically through interactions. These can become analytically and formally regulated. In the dialogical framework, schematization emerges out of the process of elementary teaching and learning. The question of the validity of certain constructions in this view arises through a shift from concrete embodied practices to the symbolically generalized reflection on it (Lorenz 1972, 121).

In other words, primary *praxis* deals with situations in which practices are introduced through concrete singular acts, whereas secondary level *praxis* deals with the symbolical use of these practices, which are based on a certain "fading off" of the primary level and make possible disagreements on meaning.

Constructivism actually shares with phenomenology the attempt to regain a "ground" Accordingly, Paul Lorenzen and Oskar Becker shared the idea that the being-in-the-world constitutes the very basis of any understanding of meaning (Mittelstrass 2002). The difference is that, in order to attain this being-in-the-world, Becker has followed an "existential-philosophical phenomenological program" ("*ein existentialphilosophisch transformiertes phänomenologisches Programm*")[36] while Lorenzen has developed the program of a *protolanguage*, on which a constructive scientific language (called *Orthosprache*[37]) (Lorenzen and Schwemmer 1975, 24) could deploy itself from its very operative bases.[38] Nevertheless this difference takes shape on a common ground which is the concern for *anthropological reality*, that is for a real synthesis principle, whose possible indication goes beyond the expressivity

[36] Perhaps, as aforementioned, Becker remains, in doing so, a Cartesian in Brandom's sense, at least from a purely semantic point of view.

[37] «Wir können alle die Redeteile, die wir durch empragmatisches Reden in ihrem Verstandnis für hinreichend gesichert halten, als unproblematisch und verständlich zu nehmen» (22).

[38] Incidentally, P. Lorenzen considered that Brouwer's own explanation of the principle of excluded-middle missed the target: "Unglüclicherweise ist die Erklärung, die Brouwer selbst für dieses Phänomen [the invalidity of "*tertium non datur*"] anbietet, eine esoterische Angelegenheit: nur, wer den Meister selber hörte, versteht ihn" (Lorenzen 1960). However, Brouwer's argument from 1908 does not presuppose any esoteric conception: it is just the fact that there is no general decision procedure for mathematical propositions.

of any language. This concern certainly has to be related to intuitionism, especially to its most controversial aspects, namely the Creating-Subject-device.

We posit that Brouwer's position aims at bringing together both, a constructability requirement, which can certainly be rendered out of the dialogical framework, *and* a hermeneutical approach to mathematical activity, which binds it to "human destiny". Such an approach must necessarily rely on *human transmission* which permeates mathematics without being of an explicitly mathematical order and which can certainly also find a place in the analytical schematization of intersubjective interactions, as long as analyticity applies, secondarily, to a domain of possible meanings, which exceeds the very domain of objectivity, validity and truth.[39]

Acknowledgements Thanks to Shahid Rahman for his endless support and to Claudio Majolino for suggesting that I read and study Oskar Becker in the first place. To Mark van Atten, who patiently answered my innumerable questions, whose writings introduced me to Brouwer's thought (making the present study possible), and who read and provided invaluable clarifications for the first version of this paper, I owe an enormous debt of gratitude. Finally, I would like to thank Christina Weiss, for having given me the opportunity to write it.

References

Bassler, O. B. (2006). Book review: Mark van Atten, On Brouwer. *Notre Dame Journal of Formal Logic, 47*(4), 581–599.

Becker, O. (1927). *Mathematische Existenz: Untersuchungen zur Logik und Ontologie mathematischer Phänomene*. Halle a. d. S.: Niemeyer.

Becker, O. (1929). Von der Hinfälligkeit des Schönen und der Abenteuerlichkeit des Künstlers. I: *Jahrbuch für Philosophie und phänomenologische Forschung. Ergänzungsband*. Halle: Husserl-Festschrift (pp. 27–52).

Becker, O. (1936). Husserl und Descartes. *Archiv für Rechts- und Sozialphilosophie, 30*, 616–621.

Becker, O. (1962). Zwei phänomenologische Betrachtungen zum Realismusproblem. In K. Hartmann (Ed.), *Lebendiger Realismus : Festschrift für Johannes Thyssen* (pp. 1–26). Bonn: H. Bouvier.

[39]It will be worth investigating how *literature* can possibly hold as a complementary *cognitive* approach to mathematics, opening a way to the indication of real-life situations from which pure theoretical attitude stand out. It was clearly a concern of Hermann Broch, notably in his novel *Die unbekannte Größe*: "der Roman der Mathematiker Richard Hieck hat sich soweit mit der Mathematik zu beschfätigen, als sie zum Kristallisationspunkt jener seelischen Urkräfte geworden ist, m.a. W. soweit sie in der Mechanik seelischen Geschehens selber Symbolwert besitzt und der Erkenntnisvorgang der Mathematik als Exponent der tieferen Seelendynamik dient" (Broch 1933). K. Lorenz, by taking up the distinction between *poiesis*, defined as "determination of what can become a sign and how", and *mimesis*, which is the successful result of research through invention, and by articulating it to the distinction between *knowledge by acquaintance* and *knowledge by description*, or between mere *exclamation* and *report* deployed certainly the bases for a dialogical treatment of that matter of the "common roots of arts and sciences". Empractical anchoring of the meaning process constitutes here the deepest ground on which a broadened constructive semantic should deploy itself beyond the limits of mere formal languages. See "On the perceptual and conceptual knowledge" in (Lorenz 2010).

Brandom, R. (2000). *Articulating reasons: An introduction to inferentialism*. Cambridge, London: Harvard University Press.

Broch, H. (1933). *Die unbekannte Größe*. Berlin: Fischer Verlag (2. Auflage 2016. Suhrkamp Taschenuch).

Brouwer, L. E. J. (1949). Consciousness, philosophy, and mathematics. In *Proceedings of the Tenth International Congress of Philosophy*, Amsterdam, August 11–18, 1948. North-Holland Publishing Company, Amsterdam 1949, pp. 1235–1249. *The Journal of Symbolic Logic, 14*(2), 1235–1249.

Brouwer, L. E. (1975). *Collected works*. In A. Heyting (Ed.),North-Holland, Amsterdam.

Bühler, K. (2011). *Theory of language: The representational function of language* (D. Fraser Goodwin, Trans.). Amsterdam, Philadelphia: John Benjamins Publishing Company.

Cavell, S. (1979). *The claim of reason*. Oxford, New York: Oxford University Press.

Dango, A. B. (2016). *Approche dialogique de la révision des croyances dans le contexte de la théorie constructive des types de Per Martin-Löf*. London: College Publication.

Descartes, R., (1999). *Oeuvres philosophiques*. (F. Alquié, Ed.) *II: 1638–1642*. Paris: Garnier.

Dummett, M. (1977). *Elements of Intuitionism*. Oxford: Clarendon Press.

Gethmann, C. F. (2002). Hermeneutische Philosophie und Logischer Intuitionismus. In J. Mittelstrass & A. Gethmann, *Die Philosophie und die Wissenschaften* (Fink). München (pp. 109–128).

Heidegger, M. (1979). *Sein und Zeit*. Tübingen: M. Niemeyer.

Heyting, A. (1931). Die intuistonnische Grundlegung der Mathematik. *Erkenntnis, 2*, 106–115.

Husserl, E. (1962). *L'origine de la géométrie* (J. Derrida, Trans.). Paris: Presses Universitaires de France.

Husserl, E. (1969). *Formal and transcendental logic* (D. Cairns, Trans.). The Hague: Martinus Nijhoff.

Husserl, E. (1973a). *Experience and judgment: Investigations in a genealogy of logic* (L. Landgrebe, Ed., S. Churchill & K. Ameriks, Trans.). London: Routledge and K. Paul.

Husserl, E. (1973b). *Zur Phänomenologie der Intersubjektivität: Texte aus dem Nachlass. Zweiter Teil: 1921–1928*. In Kern, I., & Breda, H. L. van (Ed.). The Hague: Martinus Nijhoff.

Husserl, E. (1983). *Ideas pertaining to a pure phenomenology and to a phenomenological philosophy: First book: General introduction to a pure phenomenology* (F. Kersten, Trans.). The Hague, Boston; Lancaster: Kluwer Academic Publishers.

Husserl, E. (1995). *Cartesianische Meditationen: Eine Einleitung in die Phänomenologie* (Vol. 3, duchgesehen Auflage). Hamburg: Meiner.

Husserl, E. (2001a). *Logical investigations*. Vol. 1 (J. N. Findlay, Trans.). London: Routledge.

Husserl, E. (2001b). Logical investigations. Vol. 2 (J. N. Findlay, Trans.). London: Routledge.

Kern, I. (1964). *Husserl und Kant*. Den Haag: Nijhoff.

Kreisel, G. (1967). Informal rigour and completeness proofs. In I. Lakatos (Ed.). *Studies in logic and the foundations of mathematics* (Vol. 47, pp. 138–186). Elsevier, Amsterdam.

Lorenz, K. (1972). Der dialogische Wahrheitsbegriff, *Neue Hefte für Philosophie*. H 213, pp. 111–123.

Lorenz, K. (2009). *Dialogischer Konstruktivismus*. Berlin, New York: De Gruyter.

Lorenz, K. (2010). *Logic, language and method: On polarities in human experience*. Berlin, New-York: De Gruyter.

Lorenzen, P. (1960). Logik und Agon, *Atti. Congr. Internat. De Filosofia*, vol. 4. Sansoni, Firenze, pp. 187–194. Reprinted in Lorenzen, P., & Lorenz K. (1978) *Dialogische Logik*, Darmstadt: wissenschaftl. Buchgesellschaft.

Lorenzen, P., & Schwemmer, O. (1975). *Konstruktive Logik, Ethik und Wissenschaftstheorie*. Meisenheim: Anton Hein.

Marion, M. (1998). *Wittgenstein, finitism, and the foundations of mathematics*. New York: Oxford University Press.

Martin-Löf, P. (1990). Mathematics of infinity. In *Proceedings of the International Conference on Computer Logic* (pp. 146–197). London, UK, UK: Springer-Verlag.

Martin-Löf, P. (2017). Assertion and request. *Lecture held at Oslo*, 2017. Transcription by A. Klev.

Mittelstrass, J. (2002). Oskar Becker und Paul Lorenzen oder: die Begegnung zwischen Phänomenologie und Konstrüktivismus. In J. Mittelstrass & A. Gethmann, *Die Philosophie und die Wissenschaften* (Fink). München (pp. 65–83).
Nancy, J.-L. (2000). *La pensée dérobée*. Paris: Galilée.
Narasina, R. (2009). The chequered history of epistemology and science: The intuitionist interlude. In B. Ray (Éd.), *Different types of history* (pp. 106–112). Pearson Education India.
Prabhavananda, & Isherwood, C. (Trans.). (1956). *Bhagavad-Gita: The song of God*. London: Phoenix House.
Rahman, S., Redmond, J., & Clerbout, N. (2016) N. Objective Knowledge and the not Dispensability of Epistemic Subjects. Some remarks on Popper's notion of objective knowledge. *Cahiers d'Epistémologie* (Vol. 5, pp. 25–53). L'Harmattan.
Rahman, S., McConaughey, Z., Klev, A., & Clerbout, N. (2018). *Immanent reasoning or equality in action: A plaidoyer for the play level*. Dordrecht: Springer.
Reinach, A. (2012). *The Apriori foundations of the civil law: Along with the lecture*. In J. F. Crosby (Ed.). «*Concerning phenomenology*». Frankfurt: Ontos.
Ricoeur, P. (2004). *À l'école de la phénoménologie*. Paris: J. Vrin.
Shafiei, M. (2018). *Meaning and Intentionality: A dialogical approach*. London: College Publications.
Sundholm, G. (1984). Brouwer's anticipation of the principles of charity. *Proceedings of the Aristotelian Society, 85*, 263–276.
Sundholm, G. (2014). Constructive recursive functions, church's thesis, and Brouwer's theory of the creating subject: Afterthoughts on a Parisian joint session. *Constructivity and computability in historical and philosophical perspective* (pp. 1–35). Dordrecht: Springer.
Troelstra, A. (1977). *Choice sequences: A chapter of Intuitionist Mathematics*. Oxford: Clarendon Press.
Troelstra, A., & van Dalen, D. (1988). *Constructivism in mathematics: An introduction* (Vol. 1). Amsterdam: Elsevier.
van Atten, M. (2002a). *On Brouwer*. Belmont, Calif.: Wadsworth Publishing Co Inc.
van Atten, M. (2002b). Phenomenology's reception of Brouwer's choice sequences. In V. Peckhaus (Ed.), *Oskar Becker und die Philosophie der Mathematik* (pp. 101–107). München: Fink Verlag.
van Atten, M. (2007). *Brouwer meets Husserl: On the phenomenology of choice sequences*. Netherlands: Springer.
van Atten, M. (2015). Troelstra's Paradox and Markov's Principle. In G. Alberts, L. Bergmans, & P. Muller (Eds.), *Dutch significs and early criticism of the vienna circle*. Dordrecht: Springer. (Forthcoming, Preprint on Hal, archives ouvertes).
van Atten, M. (2018). The creating subject, the Brouwer-Kripke Schema, and infinite Proofs. Forthcoming in *Indigationes Mathematica*.
van Atten, M., & van Dalen, D. (2002). Arguments for the continuity principle. *The Bulletin of Symbolic Logic, 8*(3), 329–347.
Van der Schaar, M. (2011). The cognitive act and the first-person perspective: An epistemology for constructive type theory. *Synthese, 180*, 391–417.
Wittgenstein, L. (1989). *Philosophische Bemerkungen*. In R. Rhees (Ed.). Frankfurt am Main: Suhrkamp.

Part II
Critical Positions Towards Integrating Transcendental Phenomenology and Constructivism

Chapter 5
Constitution and Construction

Mirja Hartimo

Abstract In the recent literature on Husserl's philosophy of mathematics the notions of constitution and construction have been assimilated (Da Silva 2017; van Atten 2017). The aim of this paper is to explain why this is problematic. The crux of the argument is that while construction in Husserl's texts takes place in the natural mathematical attitude, constitution of the mathematical reality is revealed in the transcendental phenomenological attitude. Equating the two notions leads to misreading either Husserl's notion of the transcendental attitude (cf. Da Silva 2017; Hartimo 2017) or else to a failure to appreciate the role of the natural attitude in Husserl's approach (cf. van Atten 2017; Hartimo 2016). Since both mathematics and phenomenology are eidetic sciences, i.e., they are both about ideal structures, the difference between the two is perhaps difficult to grasp. However, a closer look shows them to be fundamentally different, both in method and in subject matter. The most fundamental difference between construction and constitution is in the attitude: in Husserl's approach, construction takes place in natural mathematical attitude and constitution is something studied in transcendental phenomenological attitude.

Keywords Phenomenology · Logical construction · Constitution of abstract objects · Husserl

To explain all this, I will start by discussing Husserl's idea of correlation, i.e., the idea how what is experienced as objective and the manners of givenness of the objective are correlated. The purpose of this discussion is to remind the reader of the important differences between the two attitudes in general. I will then explain the differences between construction and constitution in mathematics by means of these two attitudes. After this, I will move on to discuss Husserl's notion of "construction" in more detail.

Husserl uses the term "construction" in (at least) two senses. In a wider sense, construction refers to any manner of building mathematical theories. In its narrower sense the term refers more specifically to "logical construction," that is an activity carried out in "theory of judgment". Considering the differences in the constitution

M. Hartimo (✉)
Tampere University, Tampere, Finland
e-mail: mirjahartimo@gmail.com

C. Weiss (ed.), *Constructive Semantics*, Logic, Epistemology, and the Unity of Science 44, https://doi.org/10.1007/978-3-030-21313-8_5

of the "logically constructed" versus the (mathematically) "constructed" reveals the two kinds of construction to be guided by different kinds of evidence. I will end the paper by considering the roles of logical construction and constitution in Husserl's philosophy of mathematics more generally.

5.1 Correlation

Phenomenology is a study of the correlation between the world and its subjective manners of givenness. In the *Crisis* Husserl writes that

> [t]he first breakthrough of this universal a priori of correlation between experienced object and manners of givenness (which occurred during [the] work on my *Logical Investigations* around 1898) affected me so deeply that my whole subsequent life-work has been dominated by the task of systematically elaborating on this a priori of correlation (*Crisis* 166n).[1]

And indeed, his work since 1898 can be regarded as an elaboration of the correlation, that is, how, thanks to my subjective achievements, the objective world is given to me. The correlation is between the world and our activities of structuring it. According to *Ideas I*, the objective end of the correlation is the world as encountered in the so-called natural attitude. The natural attitude is the attitude in which we find ourselves "naturally", within our experiences, in *medias res*, as Quine would have it—no attempt at looking at the world from the outside is involved in it. The world of the natural attitude is the way in which *I* take the world to be there (described from the first-person perspective): I take it as spreading endlessly in space and time. I can sense it in various ways and I can move in it in various ways. There are corporeal things and other people and I take them to be there even if I am not looking. Some things I see indistinctly and unclearly. The immediate environment is vested with expectations of what could happen. Furthermore, the "world is not there for me as a mere world of facts and affairs, but, with the same immediacy, as a world of values, a world of goods, a practical world" (Ideas I, §27). The world of natural attitude has more specified worlds that result from the more specified attitudes nested in it. Thus, for example, the world of arithmetical attitude is there for me as long as I am in the arithmetical attitude, absorbed in proving something, while the natural world in the usual sense is there for me continuously (Ideas I, §28). Furthermore, I do not think of myself to be the only one in the world. I understand other people to be similar subjects of experiences as what I am (Ideas I, §29). I also think that the world is actually there. I do not doubt its existence, nor do I doubt other humans' existence. Because of this Husserl characterizes the natural attitude by a "Generalthesis", general positing of existence.

[1] "Der erste Durchbruch dieses universalen Korrelationsapriori von Erfahrungsgegenstand und Gegebenheitsweisen (während der Ausarbeitung meiner 'Logischen Untersuchungen' ungefähr im Jahre 1898) erschütterte mich so tief, daβ seitdem meine gesamte Lebensarbeit von dieser Aufgabe einer systematischen Ausarbeitung dieses Korrelationsapriori beherrscht war." (Krisis, 169n).

5.2 Constitution

The description of the natural world sets up the task for transcendental phenomenology, the task of describing how the world is given, how we *constitute* it. The constitution refers to the achievements of our consciousness due to which we experience the world as described in the natural attitude. In the natural attitude we are unaware of the way in which the world is constituted. In order to shift our focus from our natural interests to the nature of constitution, '*phenomenological reduction*' has to be effected. This means that the general thesis, the positing of the existence of the world is bracketed or interrupted. Yet, in bracketing the naïve positing of the world's existence, the phenomenologist does not lose any of the predicates or determinations that belong to the world. The entire world remains continually there, with all its determinations, available for the phenomenologist to start examining it (Ideas I, §32). In the phenomenological reduction then,

> we put all those positings 'out of action,' we do not 'participate in them;' we direct our seizing and theoretical inquiring regard to pure consciousness in its own absolute being. That, then, is what is left as the sought-for 'phenomenological residuum,' though we have 'excluded' the whole world with all physical things, living beings, and humans ourselves included. Strictly speaking, we have not lost anything but rather have gained the whole of absolute being which, rightly understood, contains within itself, 'constitutes' within itself, all worldly transcendencies (§50, 113).[2]

In other words, the reduction changes our focus from theorems and proofs to examine the activity of constructing in pure consciousness. In pure consciousness all the worldly transcendences are, so to say, constituted. As is well known, Husserl's best-known examples are about the constitution of the objectively existing object of perception, such as a tree. We constitute the perceived object as a unity even though it is given in data that is in flux: in adumbrations with an external and internal horizon, i.e., anticipations of what else is given with the thing and what the thing will look like on the other side, etc. In *Ideas* I Husserl distinguishes between the physical thing and the *noema* that is constituted by us:

> The tree simpliciter, the physical thing belonging to Nature, is anything but (nichts weniger als) this perceived tree as perceived which, as perceptual sense, inseparably belongs to the perception. The tree simpliciter can burn up, be resolved into its chemical elements, etc. But

[2]"... setzen wir all diese Thesen, die aktuellen und im voraus die potenziellen 'außer Aktion', wir machen sie nicht mit; unseren erfassenden und theoretisch forschenden Blick richten wir auf das *reine Bewußtsein in seinem absoluten Eigensein.* Also das ist es, was als das gesuchte '*phänomenologische Residuum*' übrig bleibt, übrig, obwohl wir die ganze Welt mit allen Dingen, Lebewesen, Menschen, uns selbst inbegriffen 'ausgeschaltet' oder besser eingeklammert haben. Wir haben eigentlich nichts verloren, aber das gesamte absolute Sein gewonnen, das, recht verstanden, alle weltlichen Transzendenzen als intentionales Korroelat der ideell zu verwirklichenden und einstimmig fortzuführenden Akte habitueller Geltung in sich birgt, sie in sich 'konstitutiert'." (Ideen I, 118–119).

> the sense—the sense of this perception, something belonging necessarily to its essence—cannot burn up; it has no chemical elements, no forces, no real properties (Ideas I, 216).[3]

The distinction between the physical thing and the *noema* highlights the importance of distinguishing between the natural and the transcendental point of view. While the tree that grows belongs to the natural world, when it is regarded transcendentally it cannot burn up. Making things and ceasing from existence is only possible for the things in the natural world as well as for people as natural beings. The examination of how the tree is given, i.e., how its sense is constituted does not bring the tree into existence. In general, constitution does not bring things, whether physical or abstract, into existence. It is about the givenness, the sense of things.

5.3 "Construction" Versus Constitution

The difference between the transcendent object of the natural attitude and the examination of its givenness, i.e., the examination of the constitution of its sense is obvious when the focus is on physical objects. However, since mathematics is about abstract objects, its subject matter is ideal, and, in that respect, like the structures that can be found in pure consciousness. For this reason, both mathematics and phenomenology, in Husserl's view are *eidetic sciences*. Their objects are, in Husserl's terminology, ideal entities, essences. Husserl's term "essence" aims to capture the necessary, invariant structural features of the phenomena.

Withstanding the similarities between mathematics and phenomenology, or construction and constitution, some of their differences can be brought to light by means of two conceptual distinctions that Husserl establishes in the *Ideas I*: First, according to Husserl, the ideal essences may be either *exact* or *morphological*. In mathematics and physics they are *exact*. They can be derived deductively from the axioms, but they can also be idealized as ideal 'limits' of the free variation of imagination that cannot be found in sensuous intuition. The essences that pertain to pure consciousness are in turn morphological i.e., essentially inexact essences (Ideas I, §74).

Even more decisive is the distinction between what is *transcendent* versus that which is *immanent* to pure consciousness: To reveal the constitution of any kinds of objects one has to perform the phenomenological reductions. In the reductions, the naïve positing of the existence of nature, sciences and humanities, God, pure logic, and mathematics, as well as the material eidetic disciplines are parenthesized, i.e., interrupted (Ideas I, §59). As explained above, what is bracketed does not disappear but it can be referred to only as parenthesized, as transcendent to the pure consciousness. The pure consciousness is not empty, but it is pure from anything

[3]"Der *Baum schlechthin*, das Ding in der Natur, ist nichts weniger als dieses *Baumwahrgenommene als solches*, das als Wahrnehmungssinn zur jeweiligen Wahrnehmung unabtrennbar gehört. Der Baum schlechthin kann abbrennen, sich in seine chemischen Elemente auflösen usw. Der Sinn aber—Sinn *dieser* Wahrnehmung, ein notwendig zu ihrem Wesen Gehöriges—kann nicht abbrennen, er hat keine chemischen Elemente, keine Kräfte, keine realien Eigenschaften" (Ideen I, 222).

existing in the sense things (whether physical or abstract) exist in the world. The realm of pure consciousness is an eidetic sphere, like the world of mathematics. But whereas the essences examined in phenomenology are *immanent*, the ones in mathematics are *transcendent* (note that this is the case even if the mathematics in question is constructivist). The immanent structures clarify the *manner* of givenness of the transcendent entities. They are the ideal structures used in making sense of the transcendent world. Phenomenology and mathematics thus belong to crucially different dimensions that should not be conflated. This means that the facts from the sciences cannot be used in phenomenological explanations. As Husserl puts it, "not a single theorem, indeed not even an axiom, can be taken and admitted as a premise for phenomenological purposes…" (Ideas I, §61).[4] Consequently, the phenomenological method is bound to be descriptive as opposed to seeking for nomological explanations. But phenomenology can be, and often is, about mathematics. Then it is a study of the constitution of the concepts of theories and mathematical objects, such as numbers, that are nevertheless transcendent. Phenomenological analyses thus clarify how we constitute the concepts of exact essences (of mathematics). While the constituted concepts are usually morphological, by means of eidetic variation the "ideal" exact concepts, correlated with exact essences, can be constructed in the phenomenological attitude (Ideas I, §74). Thus Husserl hopes to be able to clarify the basic concepts used in the axiomatic theories. In his *Formal and Transcendental Logic* (1929), Husserl however claims that the mathematicians (but not applied mathematicians) need not to be concerned about how mathematics relates to intuition (FTL, §52). In that work, his constitutive analyses focus on the evidences and the presuppositions used in mathematics.

In Husserl's texts construction is constructing theories, proofs, and concepts in mathematics, in the natural mathematical attitude. When the mathematicians explain how they constructed a particular theorem, they refer to theorems, rules or strategies of proof that were used during the course of the proof. When describing the constitution of something, one *describes* the presuppositions and the evidences used in the same procedure. The revealed constitutive structures are morphological, essentially inexact, so that they do not admit axiomatization but can only be described. Examples of such structures in *Formal and Transcendental Logic* are the identity of the ideal objects and the kinds of evidence found in the mathematical construction of a theorem. Moreover, the consciousness as a whole, including the situation in which the judging subject finds herself, will be taken into account in the explanation of the constitution (FTL Appendix II, §2b). Consider, for example, the following description of constitution of a judgment:

> The judgment does not exist only in and during the active constitution, as being livingly generated in this process; rather it becomes the continuously abiding selfsame judgment, as a preserved *acquisition* dependent on functionings of passivity, these being involved everywhere in the constitution of identically persisting unities, including formations produced

[4] "The only propositions of logic to which phenomenology might ever have occasion to refer would therefore be mere logical axioms, like the law of contradiction, axioms the universal and absolute validity of which it would be able to make evident, however, on the basis of examples included among its own data" (Ideas I, §59).

actively. So far we have seen only that the acquisition, as an abiding one, is constituted, in the first place, during the living progression of retentional modification, up to the limit where the acquisition is no longer prominent.

Without this sort of preservation in a passive continuous identification, advancing judgment-processes – as a living further-forming and connecting of meant categorialia to make the unity of continually new judgments at higher and higher levels – would not be possible. The retentionally subsiding component formations remain, with this modification, within the scope of the judger's unitarily thematizing regard; he can reach back and seize them again, each as having its identical sense; also, in consequence of the new judgment-steps, they can undergo further accretions of sense in new formings (FTL, Appendix II §3a, 320).[5]

The active synthesis of a judgment does not take place in isolation but it is dependent on many other acquisitions, such as previously made judgments that are passively held in consciousness. After having been made, the judgments sink into the sedimentations of passivity. But because of the associations "a proposition, a proof, a numerical formation, or the like, can come to mind … long after the original generating has disappeared; …" (FTL, Appendix II §3b). For example, one may use a certain theorem in a proof that one has seen used somewhere else. The theorem is then passively given in one's consciousness. It is a part of the sedimentations acquired earlier—the sedimentations that persist in consciousness and that can be brought to mind when needed. Due to sedimentations one has all kinds of expectations and "hunches" about what might work and where:

When the mathematician, on the basis of the structure and sequence of formulas that he finds in his thinking-situation, anticipates a new theorem and a proof for it with an appropriate style – being guided, obviously, by association, which has obscurely awakened earlier similar situations, formulas, and combinations of formulas -, he has not yet found, as he very well knows, any actual cognition, any actual theorems of proof; and that signifies for him, as an analytic mathematician, that he has not yet made those actual judgments and combinations of judgments, in the actual activity of which each thing would spring to the fore from analytic relationships belonging to it originally. That is why he now strives for the explicit action which is his rational activity proper – no matter how necessary the associatively indicative action may remain, as pointing ahead to goals and ways for his rational practice (FTL, Appendix II, §6, 327–8).[6]

---

[5]"Das Urteil ist nicht nur in und während der aktiven Konstitution als in ihr lebendig sich erzeugendes, sondern wird zum kontinuierlich verbleibenden selben Urteil, als einem sich erhaltenden Erwerb, der eben auch für active Gebilde—wie überall (das ist in jedweder Konstitution identisch verharrender Einheiten) auf Funktionen der Passivität beruht. Soweit wir bisher gekommen sind, ist der Erwerb als bleibender zunächst nur konstitutiert während des lebendigen Fortganges der retentionalen Abwandlung bis zum Limes der Unabgehobenheit.

Diese Art der Erhaltung in passiv-kontinuierlicher Identifizierung macht allein fortschreitende Urteilsprozesse als lebendige Fortbildung und Verknüpfung kategorialer Vermeintheiten zur Einheit immer neuer und höherstufiger Urteile möglich" (FTL, 319).

[6]"Wenn der Mathematiker auf Grund des Baues und der Folge von Formeln, die er in seiner Denksituation vorfindet, nun einen neuen Satz und einen in entsprechendem Stil dafür zu führenden Beweis antizipiert—offenbar von der Assoziation, die Frühere Denksituationen, Formeln und Formelverbände dunkel geweckt hat, geleitet—so hat er, wie er wohl weiβ, noch keine wirkliche Erkenntnis, keine wirklichen Sätze und Beweise gefunden, und das besagt für ihn als Analytiker, er hat nicht die wirklichen Urteile und Urteilsverbände aktiv hergestellt, in deren wirklicher Aktivität alles aus original zugehörigen analytischen Verhältnissen hervorspringen würde. Eben darum erstrebt er nun die

In sum, the constitution of the ideal entities becomes visible only in the phenomenological attitude. While the ideal essences may be exact and transcendent, the essential structures revealed by the clarification of their constitution are inexact and can only be described. These essences are immanent to pure consciousness. They are used in the constitution of the concepts of the exact and transcendent essences of mathematics.

5.4 Logical Construction

Husserl uses the term 'construction' for various kinds of activities. In a wide sense, it refers to any mathematical activity, i.e., any construction of a proof or a theorem in mathematics. Hence, for example, in *Prolegomena* when explaining the division of labor between mathematicians and philosophers, Husserl writes that "[t]he construction of theories, the strict, methodical solution of all formal problems, will always remain the home domain of the mathematician." Here Husserl refers to the theories in algebraic approaches but also to Riemann's approach and Cantor's theory of sets (§§70–71). In *Formal and Transcendental Logic* (1929) Husserl writes that mathematics "is the realm of infinite constructions, a realm of ideal existences, not only of 'infinite' senses but also of constructional infinities" (FTL, §74). He thus seems to think that the transfinite sets are constructed. In its wide sense, construction thus refers to any mathematical theorem-proving.

However, in *Formal and Transcendental Logic* Husserl also uses the term "construction" [Konstruktion] to refer to active judging in accordance to the rules of pure apophantic logic. Husserl writes, that assuming the pure theory of grammar,

> a closed system of fundamental forms emerges, out of which, in accordance with a set of appertinent eidetic laws, ever new, ever more highly differentiated forms, and finally the system of all conceivable judgment-forms without exception, can be generated by construction [konstruktive erzeugt werden können], with the infinity of their differentiated and always-further-differentiable configurations (FTL §13b, 50).[7]

From the judgment 'S is p' one can construe the form 'Sp is q' and then '(Sp)q is r'. These judgments can be 'modified' so that they can occur as component parts in e.g., a conjunction or a hypothetical form of judgments. Such construction is law-governed and reiterative:

> *Every operative fashioning of one form out of others has its law*; and this law, in the case of operations proper, is of such a nature that the generated form can itself be submitted

---

explizite Aktion, die seine eigentliche Vernunfttätigkeit ist—wie sehr die assoziative indizierende notwendig bleibt, ihm Ziel und Wege für seine Vernunftpraxis vorzudeuten" (FTL, 325).

[7] "Um die Idee dieser reinen Formenlehre zu erfassen, hätte man sich klar machen müssen, daβ im Absehen auf eine Klassifikation möglicher Urteile überhaupt hinsichtlich ihrer Form 'Grundformen' hervorgehen, bzw. ein geschlossenes System von Grundformen, aus denen vermöge einer eigenen Wesensgesetzlichkeit immer neue, immer reicher differentzierte Formen und schlieβlich das System aller erdenklichen Urteilsformen überhaupt in der Unendlichkeit ihrer differentzierten und sich immer wieder differenzierenden Gestalten konstruktive erzeugt werden können" (FTL, 55).

> to a repetition of the same operation. *Every law of operation thus bears within itself a law of reiteration.* Conformity to this law of *reiterable operation* extends throughout the whole province of judgments, and makes it possible to construct reiteratively (by means of fundamental forms and fundamental operations, which can be laid down) the infinity of possible forms of judgments (FTL §13c, 52–3).[8]

Later on, in the same work, Husserl refers to a theory of judgments as a "transitional link" between the logic of non-contradiction and the truth-logic. In it, construction starts from judging something about an individual object. It starts from the ultimate forms of judgments that feature "ultimate subjects", "ultimate predicates", "ultimate universalities", "ultimate relations" (§82), and it builds iteratively the theory of judgments, first on the level of truths, then analogously on the level of senses. Complex judgments are then effectively reducible to the elementary judgments.[9] Husserl does not explain why the complex judgments should be reducible to the simple ones, but it seems to be crucial for a theory of judgments to be able to transfer evidence unproblematically from simple judgments of perception to the more complex ones. I will call this more limited notion "logical construction" as opposed to the more general mathematical construction. These different notions of construction show another difference between constitution and construction. Whereas there are several different kinds of construction, the task of revealing the constitution of what is constructed remains the same. Whereas the scope of what can be constructed differs depending on the kind of construction in question, everything, for Husserl, is constituted—not only mathematical objects.

While Husserl is always very clear about the dangers of modeling phenomenology after mathematics, in *Formal and Transcendental Logic* he ascribes to logical construction a function that is rather philosophical, even if it is not transcendental phenomenological. For him the theory of judgments is a vehicle that is able to transfer evidence from simple judgments to the complex judgments, and thus to greater parts of mathematics. It thus can be said to give content to at least some of the abstract objects of formal mathematics. Husserl further claims that such "reductive deliberations … uncover the hidden intentional implications included in judging and in the judgment itself as the product of judging" (FTL §85). Ultimately it.

> *yields, even for the theory of forms* and, subsequently, for procedure in *an analytics of consequence-relationships, a principle of genetic order*, which at the same time determines the specifically logical aim conferred on analytics with the concepts and laws of *truth*. With respect to the subjective, that signifies that the predelineated *order of judgment-forms involves*

[8]"*Jede operative Gestaltung einer Form aus Formen hat ihr Gesetz*, und dieses ist bei den eigentlichen Operationen von einer Art, daβ das Erzeugte abermals derselben Operation unterzogen warden kann. *Jedes Operationsgesetz trägt also in sich ein Gesetz der Iteration.* Diese Gesetzmäβigkeit *iterierbarer Operation* geht durch das ganze Urteilsgebiet hindurch und ermöglicht es, mittels aufzustellender Grundformen und Grundoperationen iterative die Undendlichkeit der möglichen Urteilsformen zu konstruieren" (FTL, 57).

[9]Husserl writes that "any actual or possible judgment leads back to ultimate cores when we follow up its syntaxes; accordingly that it is a syntactical structure built ultimately, though perhaps far from immediately, out of elementary cores, which no longer contain any syntaxes" (FTL §82, 202–3).

> *a* predelineated *order in the process of making materially evident* and in the different levels of *true materialities themselves* (emphases in the original, FTL, §85).[10]

It shows how the logically constructed region is evident in a manner that ultimately refers to the evidence of the true materialities, i.e., evidence of something perceived. Husserl continues to explain that the theory of judgments thus enables uncovering what he calls the sense-genesis of judgments. Husserl suggests that logical construction, in this strict sense, thus usefully clarifies the generation of senses within the theory of judgments. It thus helps to reveal the constitution of senses. It is thus possible to use formal (i.e., non transcendental) means for clarification of sense genesis. Thus it contributes to a more encompassing project of examining how mathematics in general is constituted.

5.5 Why Construction, Why Constitution?

For Husserl everything that is given—including objects of perception, of imagination, dreams and fiction, values and moral qualities—is constituted. For a philosopher, or a phenomenologist, the task is to uncover the ways in which these different objects and regions are constituted. In phenomenology of mathematics the particularly interesting question concerns the differences between what is constructed in the wide sense as explained above, as opposed to what is logically constructed in the narrow sense. Examining the constitution of these respective regions shows that they differ in the kind of evidence with which they are given. These evidences are ultimately the normative goals for mathematicians or logicians. Mathematicians may strive for different kinds of normative goals, which explains the existence of a plurality of approaches in mathematics: a set theorist strives for a different kind of evidence than a constructivist (for more detail, see Hartimo 2012). The task for phenomenology of mathematics is to examine these different kinds of evidences so that what is genuine can be distinguished from the spurious. The ultimate purpose of giving transcendental constitutive analyses is to clarify the used concepts so that there are no paradoxes or confusions about senses and evidences. Its task is to describe our consciousness so as to "know thyself", to be clear about our thinking, about what we are doing, what we are striving for, and whether the used concepts are clear enough. The transcendental constitution analysis thus adds to the naively carried out mathematics a kind of metareflection that is needed for it to be "knowingly" carried out. The intuitionist (in contrast to some constructivists), regarded from the Husserlian point of view, combines this kind of metareflection to his acts of construction

[10]"Hieraus ergibt sich *schon für die Formenlehre* und dann für das Vorgehen in einer *Analytik der Konsequenz* ein *Prinzip genetischer Ordnung*, die zugleich bestimmend wird für das spezifisch logische Absehen der Analytik, das mit den Wahrheitsbegriffen und Sätzen zum Austrage kommt. In subjektiver Hinsicht besagt das, daβ die vorgezeichnete *Ordnung der Urteilsformen zugleich in sich birgt* eine vorgezeichnete *Ordnung sachlicher Evidentmachung* und in der Abstufung der *wahren Sachlichkeiten selbst*" (FTL, 215).

without distinguishing between the two points of view. Regarded from this point of view, the intuitionist is not naïve, but restricted to one set of epistemic norms. In addition, Husserl thinks that this kind of metareflection should likewise be added to the other ways of doing mathematics. The classical mathematicians however are typically naïve "technicians", as Husserl would put it. The advantage of the transcendental point of view is that in it different kinds of attitudes of doing mathematics together with different kinds of epistemic norms, determining what is genuine, can and should be distinguished from each other and compared to each other (cf. *Ideas II*, §49 d).

5.6 Conclusion

The main difference between constitution and construction is that construction is mathematical activity, something carried out in natural (mathematical) attitude, while examination of the constitution of the world of mathematics requires the transcendental point of view. In both of its senses, the wide sense that covers all mathematical theorem-proving and the narrow sense (logical construction in theory of judgments), construction has a "narrower" scope than constitution: everything is constituted, but only some of it is or can be constructed. For example, the trees are not constructed, even though we constitute them.

Constitution does not only have a greater scope, but it has more depth too: e.g., clarifying the constitution of a logical construction has to take into account everything in consciousness that relates to the act and the object of construction. The transcendental point of view to construction thus encompasses not only what Husserl calls active syntheses but also passive sedimentations that function in the background of active judging, and in that sense the description of the constitution of a construction makes explicit what is only implicit in construction. Curiously logical construction can be helpful for the more encompassing examination of evidences, and hence for the constitutive study. The role of logical construction (in the strict sense) is to transfer and impart evidence of the simple judgments of perception to logically more complex judgments and ultimately to at least part of mathematics, thus it reveals the sense genesis of the judgment in question. Generally, the study of constitution is important for it clarifies the presuppositions, goals, and concepts of mathematical activity, and hence renders it fully understood. Thanks to it, the phenomenologist is able to compare and contrast different kinds of mathematical practices and the evidences related to them.

References

Da Silva, J. J. (2017). *Mathematics and its applications, a transcendental-idealist perspective*. Berlin: Springer.

Hartimo, M. (2012). Husserl's pluralistic phenomenology of mathematics. *Philosophia Mathematica, 20*(1), 86–110.
Hartimo, M. (2016). Review of M. Van Atten. Essays on Gödel's reception of Leibniz, Husserl, and Brouwer. *Journal for the History and Philosophy of Logic*. 2015. https://doi.org/10.1080/01445340.2015.
Hartimo, M. (2017). Review of Jairo José da Silva. *Mathematics and its applications: A transcendental-idealist perspective,* Springer, *Notre Dame Philosophical Reviews.*
Husserl, E. (Crisis). (1970). *Die Krisis der europäischen Wissenschaften und die transzendentale Phänomenologie: Eine Einleitung in die phänomenologische Philosophie*. Hrsg. W. Biemel. Hua VI. The Hague: Nijhoff, 1954. Reprinted 1976. [English translation: *The Crisis of the European Sciences and the Transcendental Phenomenology: An Introduction to Phenomenological Philosophy*. Trans. David Carr]. Evanston: Northwestern University Press.
Husserl, E. (FTL). (1969). *Formale und transzendentale Logik.* Hrsg. Paul Janssen. Hua XVII. Den Haag: Martinus Nijhoff, 1974. [English translation: *Formal and Transcendental Logic*] Trans. D. Cairns. The Hague: Nijhoff.
Husserl, E. (Ideen I). (1983). *Ideen zu einer reinen Phänomenologie und phäneomenologischen Philosophie. Erstes Buch. Allgemeine Einführung in die reine Phänomenologie.* Herausgegeben von Walter Biemel. Husserliana Band III. Haag: Martinus Nijhoff 1950. [English translation: *Ideas pertaining to a pure phenomenology and to a phenomenological philosophy. First book, General introduction to a pure phenomenology*]. The Hague, Boston, Lancaster: Martinus Nijhoff.
Husserl, E. (Ideen II). (1989). Ideen zu einer reinen Phänomenologie und phänomenologischen Philosophie. Zweites Buch: Phänomenologische Untersuchungen zur Konstitution. Husserliana Band IV. Hrsg. W. Biemel. The Hague: Marinus Nijhoff 1952. [English translation: *Ideas pertaining to a pure phenomenology and to a phenomenological philosophy. Second book, Studies in the phenomenology of constitution. Translated by Richard Rojcewicz and André Schuwer*]. Dordrecht, Boston, London: Kluwer.
van Atten, M. (2017). Construction and constitution in mathematics. In S. Centrone (Ed.), *Essays on Husserl's logic and philosophy of mathematics*, *Synthese Library* (pp. 265–316). Springer.

Chapter 6
Husserl's Purely Logical Chastity Belt

Claire Ortiz Hill

Abstract What follows is about Husserl, whose phenomenology, I believe, can be understood as a form of constructivism. However, I generally write about Husserl's philosophy of logic and mathematics, which he repeatedly said had nothing to do with transcendental phenomenology. So, my aim here is to discuss some things that I believe people interested in constructivism and Husserl's transcendental phenomenology need to keep in mind. I say that, despite appearances, phenomenology was not everything for Husserl. As much as he loved it, he placed definite limits on what one should do with it and believed that it required an objective complement in the form of pure logic, that it had to be subject to a priori laws to keep phenomenologists from falling into psychologism, naturalism, empiricism, relativism and associated evils. According to this interpretation, Husserl the possible constructivist, Husserl the phenomenologist, Husserl the Platonist, Husserl the realist, Husserl the idealist were one and the same person from the late 1890s until his death, something which is particularly well expressed in the volumes of his lecture courses published since the 1980s, which shed considerable light on his thought.

Keywords Edmund Husserl · Pure logic · Transcendental phenomenology · Psychologism · Herman Weyl · Georg Misch · David Hilbert · Objectivity · Ideality · Karl Weierstrass · Georg Cantor · Leonard Nelson · Plato

What I have to say here is part of a broader project to show that as a result of the excitement, both positive and negative, generated by phenomenology, Husserl's pioneering theories about formal logic were never furthered as they should have been and that once they are pieced back together, philosophers can, and should, experiment with them as an alternative to the Fregeo-Russello-Quineo logic (which I call FRQL, pronounced freakl) that has propped up Analytic philosophy (which I call FRQ, pronounced freak), which unsuccessfully tried to wipe out precisely what Husserl believed was needed to uphold truly scientific knowledge.

C. O. Hill (✉)
Paris, France
e-mail: claire.hill@orange.fr

C. Weiss (ed.), *Constructive Semantics*, Logic, Epistemology, and the Unity of Science 44, https://doi.org/10.1007/978-3-030-21313-8_6

6.1 Husserl in Love

According to the still too popular account of the evolution of Husserl's thought, after a brief romance with anti-psychologistic, Platonic realism following his divorce from empirical psychology during the 1890s, Husserl was unable to resist the charms of subjectivity, espoused psychologism in a different dress and fathered transcendental phenomenology.

There is little that is true in that account. Phenomenology grew out of Husserl's troubling encounters in the 1890s with what he once called the incomprehensibly strange worlds of actual consciousness and the purely logical (Husserl 1994, 491–92; Hill 2015). He did fall out of love with empirical psychology and he did fall in love with transcendental subjectivity, but he was not the lopsided philosopher generally thought by followers and foes alike to have spent most of his life cogitating only about subjectivity. His phenomenology is not, as is often thought by both those have embraced it and those who have spurned it, in the least an autonomous science of subjectivity.

Stung by his experience with empirical psychology, Husserl spent the rest of his life fighting to avoid naïve psychologizing. He did ultimately throw himself wholeheartedly into really extensive investigations of the realm of transcendental subjectivity, of which he was unmistakably enamored, but his disappointment with empirical psychology had impressed upon him the need to expose the ultimate, objective, a priori, ideal underpinnings of science. So, while he was discovering and courting phenomenology, he devised a strategy to help those enticed by transcendental subjectivity to show self-control and refrain from committing the psychologizing and relativizing sins, the near occasion of which he wanted to make sure was avoided. And he remained faithful to those theories about objective realities until parted from them by death.

What is true about the popular account is that, enthralled by transcendental phenomenology, after a certain point, Husserl no longer desired to pursue pure logic. In 1917, he wrote to Hermann Weyl that despite all the work he had devoted to it, he had not pursued it completely to the end, because it had had to be more important to him to develop his ideas about transcendental phenomenology. In 1930, he confessed to Georg Misch that he had lost all interest in formal logic and all real ontology in the face of a systematic grounding of a theory of transcendental subjectivity (Husserl 1917/18, XXIII, nn. 1, 4). This does not mean, however, that he cast off his purely logical chastity belt or tore up his purely logical safety net. He left them in place, but after a certain point, it was transcendental phenomenology that busied his mind.

6.2 The Wedding of Formal and Transcendental Logic

The relationship that Husserl saw between transcendental phenomenology and the strictures of the pure logic that was to keep phenomenologists from falling into the temptation to go too far with his science of subjectivity is not well understood and its implications have barely been explored.

One of the main reasons why it has gone all but unstudied is that it is not sufficiently appreciated that for Husserl logic had two sides. Readers of his late work, *Formal and Transcendental Logic*, find him still stressing that logic turns *both* towards the deeply hidden subjective forms in which reason does its work *and* towards the objective order, a world of concepts, ideal objects, where truth is an analysis of essences or concepts and knowing subjects and the material world play no role (Husserl 1929, §8), that subjective, transcendental logic had to find its complement in pure, objective, a priori formal logic, free from acts, subjects, or empirical persons or objects belonging to actual reality and entirely grounded in conceptual essentialities, and vice versa. His search to comprehend the intercourse between subjectivity and objectivity was at the heart of the dynamic that brought phenomenology into being.

Science, in the objective sense, he taught, is a web of theories, and so of proofs, propositions, inferences, concepts, meanings, not of experiences (Husserl 1906/07b, §§11, 17, 19a, b). Pure logic, is the science of concepts and relations of concepts, of propositions and relations of propositions, of the possible forms grounded in these concepts and propositions. To further their insight into the essence of pure logic, Husserl once asked students to reflect on the following:

> Scientific reasoning aims for truth. Truth is realized subjectively in judgment and is stated in statements.... *Every scientific theory is a system of statements*.... It is something complete in its own right and, as it is, lays claim to truth and falsehood. The starting propositions lay claim to this directly. The theory, the definitely formed web of propositions, lays claim to substantiating new truth indirectly, step by step. And the system itself lays claim to being true as a system. That means that everywhere one thing is linked to another by logical inference that is also stated, therefore, is also set down as true (Husserl 1906/07b, §11).

He maintained that the theoretical system of modern pure mathematics was no more than a system of logically combined statement meanings, a system of propositions stating truths about a certain combination of the mathematical facts making up the field of mathematics (Husserl 1906/07b, §§11, 19c).

Husserl underscored the primacy of the objective side of logic. Pure logic, he taught, embraces all the concepts and propositions without which science would not be possible, would not have any sense or validity (Husserl 1902/03a, 47). It supplies the standards by which to measure the extent to which any presumed science meets the criteria of being a genuine science, the extent to which the particular findings of that science constitute genuine knowledge, the extent to which the methods it uses are genuine ones (Husserl 1929, §7). The world constituted by transcendental subjectivity, he insisted, is a pre-given world, a world determined and determinable in itself with exactitude, a world within which any individual entity is given beforehand as in principle determinable in accordance with the methods of exact science as being a world in itself in a sense originally deriving from the achievements of the physico-mathematical sciences of nature (Husserl 1939, §11).

Now, having said that, the underlying, undying paradox of Husserl's phenomenology remains that, while he taught that the ultimate meaning and source of all objectivity making it possible for thinking to reach beyond contingent, subjective, human acts and lay hold of objective being-in-itself was to be found in ideality and the ideal laws defining it (Husserl 1902/03a, 200), he knew that his theories about the "two-sidedness of everything logical, in consequence of which the problem-groups become separated and again combined", involve "extraordinary difficulties". He saw that since, according to his theories, the ideal, objective, dimension of logic and the actively constituting, subjective dimension interrelate and overlap, or exist side by side, logical phenomena seem to be suspended between subjectivity and objectivity in a confused way (Husserl 1929, §26c). He even suggested that almost everything concerning the fundamental meaning of logic, the problems it deals with and its method, was laden with misunderstandings owing to the fact that objectivity arises out of subjective activity. He said that even the ideal objectivity of logical structures and a priori nature of logical doctrines especially pertaining to this objectivity, and the meaning of this a priori were afflicted with a lack of clarity, since what is ideal appears as located in the subjective sphere and arises from it. He further considered that it was due to these difficulties that, after centuries and centuries, logic had not attained the secure path of rational development (Husserl 1929, §8).

So it is that Husserl's great friend and colleague, the mathematician David Hilbert, could write of how Husserl had adopted an a priori method and rejected psychologism. But, Hilbert concluded, Husserl's method was in fact psychological (Hilbert n.d., 386).

6.3 Respecting the Rules of Pure Logic

Husserl considered that the realm of truth was no disorderly hodgepodge, that truths are connected in systematic ways, governed by consistent laws and theories, that inquiry into truth and its exposition must be systematic and the systematic representation of knowledge must reflect the systematic representation grounded in the things themselves. All invention and discovery, he taught, involves formal patterns, without which there is no testing of given propositions and proofs, no methodical construction of new proofs, theories and whole systems. No blind omnipotent power has heaped together some pile of propositions P, Q, R, strung them together with a proposition S, and then organized the human mind so that the knowledge of the truth of P must unfailingly entail knowledge of S. Not blind chance, but reason and the order of governing laws reign in argumentation (Husserl 1896, 9, 13, 16–17; Hill 2018).

According to him, the pure truths of logic were all the ideal laws entirely grounded in the meaning of the concepts that all science had inherited, that is to say, in the meaning, essence or content of the concepts of truth, proposition, object, property, relation, combination, law, fact etc. Such laws must not be violated, not because that would conflict with some truth, but because it would produce *Widersinnigkeiten*, contradictions (Husserl 1900–01, *Prolegomena* §37).

So, any assertion whose content is at odds with principles rooted in the meaning of truth as such would be self-cancelling or logically *widersinnig*, its particular content in contradiction with what is rooted in the general meaning of its own meaning categories (Husserl 1900–01, *Prolegomena* §37). As an example, he gave the proposition, "Of two contradictory propositions, one is true and one false", which he said is to be viewed as absolutely certain, as simply an "unfolding" of the content of its "concepts", in which it is purely grounded. It cannot be denied without flying in the face of the meaning of those words. Anyone denying it does not know what contradictory means, what true and false mean (Husserl 1906/07b, §13c).

He saw every concrete meaning as being a fitting together of matter and form in conformity with an ideal pattern which could be set forth in formal purity and to which an a priori law of meaning corresponded that governed the formation of coherent meanings out of syntactical materials falling under definite categories having an a priori place in the realm of meanings. So, meanings were governed by a priori laws that regulate the ways in which they could be combined with new meanings, the ways in which they could fit together and constitute a meaningful, coherent whole. They could only do so in antecedently definite ways, while other possibilities of combination were excluded by laws and yielded only a heap of meanings, never a single meaning. For him, this impossibility of combining meanings in certain ways was by no means merely subjective. It was objective, ideal and grounded in the nature, the pure essence, of the realm of meaning. It was an a priori impossibility (Husserl 1900–01, IV §10).

One can freely exchange expressions within a given category, he pointed out. This is true of all meanings whatsoever. The coherent, meaningful expression 'this tree is green' can be formalized to obtain the corresponding pure meaning form 'this *S* is *p*', and so formalized, it can be interpreted in infinitely many ways. Any noun or noun phrase can be put in the place of '*S*', any adjective in the place of '*p*', and a coherent, meaningful meaning and independent proposition of the indicated form will result. Such free exchange of expressions within a given category might yield false, dumb or funny meanings, but it will necessarily yield coherent meanings. On the other hand, mere combinations of words like 'a round or', 'king but or', 'a man and is', 'this reckless is green', 'more intensive is round', 'this house is equal' are nonsensical, meaningless, utterly incomprehensible. Substituting the noun 'horse' for the relation word 'similar' in the form '*a* is similar to *b*' yields only a sequence of words in which each word has a meaning. It is completely obvious that so combined no meaning exists, or can possibly exist, for them. They break the laws about what can be meaningful. Meaning itself is missing (Husserl 1900–01, IV §§10, 12).

6.4 Abstaining from Unlawful Intercourse with Phenomenology

In order to avoid ascribing ideas to Husserl that he did not hold—and against which he strenuously militated–it is very important to realize that Husserl consistently and explicitly stated what one could not do with phenomenology (Hill 2015).

The special interest of transcendental phenomenology, he stressed, does not lie in the theoretical concepts and laws to which the sciences are subject. It does not aim at objective being and laying down truths for objective being. It is not an objective science. "What is objective belongs precisely to objective science, and what objective science still lacks for completion is its affair to obtain and its alone. *Transcendental interest, the interest of transcendental phenomenology aims at consciousness as consciousness of objects*" (Husserl 1906/07b, 432). It has "no dealings with a priori ontology, none with formal logic and formal mathematics. It is phenomenology of the constituting consciousness, and so not a single objective axiom, meaning one relating to objects that are not consciousness, belongs in it, no a priori proposition as truth for objects, as something belonging in the objective science of these objects, or of objects in general in formal universality" (Husserl 1906/07b, 432).

Now, in his paper on "Construction and Constitution in Mathematics", Mark van Atten said that as a claim about Husserl's conception of transcendental phenomenology, what I have just said, and which I said in my paper "Husserl on Axiomatization and Arithmetic" (Hill 2010, 62; Hill & da Silva 2013, 105-06) cannot be right (van Atten 2010, 48). However, it is my translation of Husserl's own words in *Einleitung in die Logik und Erkenntnistheorie* (Husserl 1906/07a), my translation of which (Husserl 1906/07b) had not yet been published at the time I prepared the paper. The reference is to the German text, which reads,

> In dieser transzendentalen Phänomenologie, haben wir es nun nicht zu tun mit apriorischer Ontologie, nicht mit formaler Logik und formaler Mathematik....
>
> Die transzendentale Phänomenologie ist Phänomenologie des konstituierenden Bewusstseins, und somit gehört kein einziges objektives Axiom (bezüglich auf Gegenstände, die nicht Bewusstsein sind) in sie hinein. In sie gehört kein apriorischer Satz als Wahrheit für Gegenstände, als welcher er hineingehört in die objektive Wissenschaft von diesen Gegenständen oder von Gegenständen überhaupt in formaler Allgemeinheit (Husserl 1906/07a, 425).

Husserl's lists of what was purely logical–therefore, not to be approached by phenomenology–included, the theory of gravity, the system of analytic mechanics, the mechanical theory of heat, the theory of metric or projective geometry, which are, he maintained, all entirely made up, not of mental experiences of one person or another, or of states of mind, but of ideal material, of meanings (Husserl 1906/07b, §12). Not a matter for phenomenology were also the basic concepts of mathematics, the theory of cardinal numbers, the theory of ordinals, set theory, mathematical physics, pure geometry, geometry as a priori theory of space, the axioms of geometry as a theory of the essences of shapes of spatial objects, the pure theory of meaning and being, a priori ontology, a priori real ontology of any kind (thing, change etc.), natural scientific ontology, objective axioms relating to a priori propositions as truths for objects, as something belonging in the objective science of these objects, essence-propositions about objects insofar as they are objective truths and as truths have their place in a truth-system in general (Husserl 1906/07b, §§18–19; Husserl 1917/18, Chap. 11). He stressed that, if

> we are not interested in the transcendental task and we remain in pure theory of meaning and being, then we practice logic, natural scientific ontology, pure theory of space, etc. These

> need not concern themselves at all with cognitive formations, with consciousness. Likewise, if we practice ethics as pure ethics (or logic of morality), esthetics, or logic of esthetic appreciation, axiology or pure logic of values.... the logic of the ideal state or of the ideal world government as a system of cooperating ideal nation states (or the science of the ideal state).... the ideal of a valuable existence (ideally valuing and valuable human beings aimed at an ideally valuable nature accommodating their values)... and the logic of this ideal... ideal-esthetic existence, pure esthetics.... Ontology of nature, ontology of minds, ontology of ethical personalities, ontology of values, etc. (Husserl 1906/07b, 434–35).

However, he went on to affirm that "*belonging to all of them are transcendental phenomenologies*" (Husserl's emphasis) that transcendentally investigate the valid objects of different categories, the objects of these ontologies, in relationship to types of consciousness essentially belonging to them (Husserl 1906/07b, 435). Indeed, though he believed that only certain of the most general cognitive-formations enter the picture for purposes of phenomenological elucidation in the case of pure logic, of an 'analytics' in the broadest, radical sense of the word (Husserl 1913, 31), he realized that even "the most trivial analytical knowledge presents big problems and hard problems for critique of knowledge. A puzzle is already present in them: How objectively valid knowledge, knowledge of things existing on their own, is possible vis-à-vis the subjectivity of knowing as a subjective activity" (Husserl 1906/07b, 335).

And, he recognized that although all fields of *theoretical* knowledge have a systemic form that belongs to formal logic itself, not all sciences are theoretical disciplines that, like mathematical physics, set theory, pure geometry or pure arithmetic, are characterized by the fact that their systemic principle is a purely analytical one based on the meanings of their concepts. For him, sciences like psychology, history, the critique of reason and, notably, phenomenology, were not purely logical and so obliged philosophers to go beyond the analytico-logical model. When those not purely logical sciences were formalized and philosophers asked what binds the propositional forms into a single system form, they faced nothing more than the empty general truth that there is an infinite number of propositions connected in objective ways that are compatible with one another in that they do not contradict one another analytically (Husserl 1917/18, §54; Husserl 1929, §35a; Husserl 1902/03b, 31–43, 49).

6.5 For Example, Pure Arithmetic

Husserl was a mathematician by training and long kept company with the most outstanding mathematicians of his time, namely Karl Weierstrass, Georg Cantor, and Hilbert and his circle (Hill 2002). So he often used examples from mathematics to illustrate the points he wanted to make.

He insisted that mathematical propositions were a priori, that all truth was nothing other than the analysis of essences or concepts (Husserl 1906/07b, §13c), that the original mathematical disciplines were disciplines of the purely logical sphere that had proceeded from given, purely logical basic concepts and axioms and directly perspicuous laws grounded in the essence of purely logical categories and had thus yielded the concept of cardinal number, the primitive laws of number given as directly perspicuous truths, and the dependent laws of number based on them (Husserl 1906/07b, §19d). He taught that, in his words (and my translation),

> If one ascends to a purely theoretical and a priori discipline.... to all the concepts that determine the objective meaning of science in general and are inseparable from it, then it is clear without further ado that all of pure mathematics belongs in this sphere, that all purely mathematical disciplines... are encompassed by pure logic as naturally conceived.... all concepts belong in pure logic that are not to be assigned to a particular science limited to particular domains of objects, but to all sciences in general and are necessarily common to them, all concepts, therefore, that have this reference to objects in general in the most universal way.... The concept of cardinal number is such a concept, and every numerically determined cardinal number belongs among these concepts. One is something in general. Anything, no matter what it is, can be posited as one.... And all numbers are built upon units.... (Husserl 1902/03b, 35).

Pure arithmetic, he held, explores what is grounded in the essence of number. It has nothing at all to do with nature. It is not concerned with physical things, souls, real occurrences of a physical or mental nature, does not acquire its universal propositions by perception and empirical generalizations on the basis of the perception and the substantiation of the resulting individual judgments. Insofar as they are really purely mathematical, all mathematical propositions express something about the essence of what is mathematical, about the meaning of what belongs to it. Their denial is consequently an absurdity. When I say that 2×2 is not 4, but 5, I am saying something "unthinkable", absurd, something nullifying the meaning of the words. Mathematicians do not state $a + 1 = 1 + a$ as a hypothesis to be established as true in further experience or inductively in conformity with the methods of the natural sciences. Rather, they start with $a + 1 = 1 + a$ as something unconditionally valid and certain, for it is obviously part of the meaning of the term "cardinal number" that each thing can be increased by one. To say that a cardinal number cannot be increased amounts to being in conflict with the meaning of "cardinal number". It amounts to not knowing what one is talking about (Husserl 1906/07b, §13c).

However, he acknowledged that, even though ordinary arithmetic, in both its naïve and its technical forms, does not at first have any common cause with theory of knowledge and phenomenology, if it undergoes phenomenological elucidation, and so learns from the sources of phenomenology to solve the great riddles arising from the correlation between pure logic and actual consciousness, and if in so doing it also learns the ultimate formulation of the meaning of concepts and propositions that only phenomenology can provide, then it will have transformed itself into truly philosophical pure logic that is more than a mere coupling of natural-objective *mathesis* with phenomenology of knowledge, but rather is an application of the latter to the former (Husserl 1913, 29–30).

6.6 Chastity as Personal Integrity, Purity in Conduct and Intention

Besides abstention from unlawful intercourse with, say phenomenology, chastity can be defined as personal integrity and purity in conduct and intention. For Husserl, his commitment to the ideal realm was a matter of integrity. He knew that what he was claiming about it was "very hotly combated as being mysticism and scholasticism" (Husserl 1917/18, §4). Indeed, on December 29, 1916, Göttingen philosopher Leonard Nelson wrote to Hilbert that Husserl,

> admittedly also originally came from the mathematical school, but... bit by bit turned more and more away from it and turned towards a school of mystical vision, whereby he also deadened the feel in his school for the demands and value of a specifically scientific method. He even goes so far, after his own lack of success with it, as to see a danger in methodological thinking and thinks that it would ruin philosophers for whom the truth only reveals itself in mystical vision. Even though Husserl himself remains protected by certain inhibitions from mystical degeneracy by virtue of strong ties to mathematics that he has not been able to cast off, one must unfortunately nonetheless note with horror that after the school as such had torn down the bridges to mathematics behind them, how unrestrainedly his students lapsed into every excess of Neo-platonic mysticism, the prevalence of which is all the more dangerous since Husserl's scientific past will unjustifiably carry over to the school the assurance that it is willing to do scientific philosophy in earnest and is capable of this. Here I fundamentally part company with Husserl's circle. For me philosophy is not a matter of mystical vision, but one of the most sober, driest thinking... (Nelson 1916, 390–391).

Husserl, though, wanted it understood that he was "far from any mystico-metaphysical exploitation of 'Ideas', ideal possibilities and such" (Husserl 1905) and he considered this a matter of integrity. He did not believe that he was according the word 'idea' any sort of mystical meaning. He taught that, as regards its essential, theoretical makeup, science is a system of ideal meanings (Husserl 1906/07b, §12), but that ideal entities had not been artificially devised by him or anyone else; they were given beforehand by the meaning of the universal talk of propositions and truths indispensable in all the sciences. This indubitable fact, he stressed, was the starting point of all logic (Husserl 1908/09, 45).

He defended himself saying that he had not embraced Ideas (in his own sense of *Ideen*) and classes of idealities because he prided himself on his own intellectual intuition and wanted to penetrate into a mystical transcendental world by means of it, or because he was looking down his nose at vile empiricism and psychologism, or because they made it easier for him to devise a "nobler" sort of philosophy that brought him a reputation of being a noble soul. He said he embraced them for the same mundane reason that he embraced things, just because he saw them and in looking at them grasped them himself (Husserl 1917/18, §8).

The world of the purely logical, he explained, is a world of ideal objects, a world of concepts (Husserl 1906/07b, §13c). The constant talk of propositions, of true and false means something identical and atemporal. No more is meant by ideality than that it is a matter of a kind of possible objects of knowledge, whose particular characteristics can, and in scientific investigation must, be determined, while they are just not objects in the sense of real objects (Husserl 1908/09, 47). He considered the recognition of ideal objects, or Ideas, with a capital I, as atemporal, supraempirical objects to be the pivotal point of all theory of knowledge and decisive for all further considerations. He taught that it was imperative for people to concede once and for all that they are genuine, actual objects, new kinds of objects (Husserl 1917/18, §8).

Ideal objects, he said, were not anything particularly lofty, but were what was the very most ordinary, like ordinary stones on the road. All people know them in a certain naïve way since they talk of numbers and do so in ideal ways. Only philosophers do not wish to know them. They dismiss them as Platonic Ideas. However, he argued, if one recognizes givens like the series of natural numbers as objectivities, one can only describe them in the way Plato did in his theory of Ideas, as eternal, selfsame, non-temporal and non-spatial, unmoved, unchangeable, etc. But, he complained, people trained in traditional philosophy then instantly think Platonic Ideas. Such Platonic realism becomes associated with mysticism and Neo-Platonism with a magical view of nature as far removed as can be from genuine natural science. Philosophers recall how this merged into scholastic realism during the Middle Ages. Anyone advocating giving ideal objects their due is open to charges of being a reactionary, mystic, scholastic, the latter two being the strongest scientific terms of abuse of his time, in which formal logic was vilified as being empty scholasticism, and espousing idealism for a pure logic undefended (Husserl 1917/18, §§5a, 19a).

He said that if someone wants to know what Ideas are, one need only point to the self-evident givens like the cardinal number series, or to absolutely self-evident statements about members of the number series. He stressed that one knows what it is being talked about when one speaks of the numbers 1, 2, 3. One knows that one is not talking about nothing, but always about an unreal-existing something.... It is absolutely clear that 2 is $1 + 1$. One thereby grasps it, has it itself, so that any doubt as to its existence would be entirely absurd. The important thing is that it is something to be had, something one can grasp, and about which self-evident statements can be made. It is an absolute, indubitable truth that $2 < 3$, that in the cardinal number series, 2 has its place between 1 and 3, etc. If absolute truths hold for 2, then 2 is precisely something as subject of these truths themselves, or it is an object, because that implies absolutely the same thing. Consequently, there are objects of insightful givenness that are not things and not existential moments in the spatio-temporal world, but are Ideas or ideal objects. Each time, one makes a self-evident statement about numbers of the number series and thus grasps an objectively valid truth, precisely these numbers, and not something else, are the objects to which the truth refers. They are therefore objects. They are not things and not moments of things. They do not exist in space and time, etc. They are precisely what they give themselves to be (Husserl 1917/18, §8).

Each cardinal number occurs only once in the number series. Infinitely many possible empirical sets can be counted: cardinal numbers of horses, of carrots, etc., but those empirical cardinal numbers come into being and pass away, start and stop, etc. That does not, however, affect the pure cardinal numbers. If no concrete cardinal number n existed in the real world from a certain point in time on, then that would not mean there was a hole in the pure number series between $n-1$ and $n+1$ (Husserl

1917/18, §8). The number 2 is not an object of perception and experience. Two apples come into being and pass away, have a place and time, but if they are eaten up, the number 2 is not eaten up. The number series of pure arithmetic has not suddenly then acquired a hole, as if we were to have to count 1, 3, 4 (Husserl 1906/07b, 13c).

To give another, non-mathematical, example, he speculated that belonging to individual human beings and the individual mind is an a priori Idea implying a whole wealth of a priori Ideas regarding all the multiple forms of consciousness and their interweaving, as well as the pure Ideas of ego, personality, character traits etc. A priori, one can surely say that a subject is conceivable without relation to other subjects. But, if we conceive of it as it really is experientially, woven into the societal context, and a co-bearer of the societal consciousness and its cultural correlates, then we see that new Ideas arise, above all, the Idea of the collective mind with all related Ideas. So, a field of a priori considerations opens up. An ontology of the collective mind, an a priori essence-theory prior to all the empirical human sciences precedes here, the way the a priori of nature does the natural sciences. This essence-theory builds itself up above the essence-theory of the individual mind, is therefore not independent of the former—just as the collective mind is a higher-order objectivity grounded in the objectivities we call individual minds. But, he saw this Idea of an a priori analysis of the collective life of the mind and its objective correlates as being so far removed from the thoughts of sociologists that they were liable to dismiss the mere suggestion that any such thing might exist and might provide the necessary epistemological basis of all genuine social science as smelling of mysticism or scholasticism (Husserl 1917/18, §64).

Husserl defended himself against such charges by saying that he was requiring nothing more than the intellectual integrity to allow the things that are prior to any theory because they are the most evident of all evident facts to count as being precisely what they proclaim themselves to be (Husserl 1917/18, §4). He said that if this was enough to have him called a scholastic, then that was all fine and good. He asked whether it was preferable to have integrity and be called a scholastic or to lack integrity and be a modern empiricist. He said that he was pleading in favor of integrity and did not fall flat on his face when he was called a scholastic. "Integrity stands the test of time", he reminded (Husserl 1917/18, §8).

References

Hilbert, D. n.d. Extracts from Hilbert's *Denkschrift* for Leonard Nelson. In Hill & da Silva 2013, 386–387.

Hill, C. O. (2002). On Husserl's mathematical apprenticeship and philosophy of mathematics. In A-T. Tymieniecka (Ed.), *Phenomenology world-wide* (pp. 76–92). Dordrecht: Kluwer. Anthologized in Hill & da Silva 2013, 1–30.

Hill, C. O. (2010). Husserl on axiomatization and arithmetic. In M. Hartimo (Ed.), *Phenomenology and mathematics* (pp. 47–71). Dordrecht: Springer. Anthologized in Hill & da Silva 2013, 93–114.

Hill, C. O. (2015). The strange worlds of actual consciousness and the purely logical. *New Yearbook for Phenomenology and Phenomenological Philosophy, 13,* 62–83.

Hill, C. O. (2018). Limning the true and ultimate structure of reality. In P. Hackett (Ed.), *Mereologies, ontologies, and facets: The categorical structure of reality*. Lanham MD: Lexington Books, 37–64.
Hill, C. O. & Jairo José da Silva (2013). *The road not taken, on Husserl's philosophy of logic and mathematics*. London: College Publications.
Husserl, E. (1896). *Logik, Vorlesung 1896*. Dordrecht: Kluwer (2001).
Husserl, E. (1900-01). *Logical investigations* 1900–01. New York: Humanities Press (1970).
Husserl, E. (1902/03a). *Allgemeine Erkennthistheorie, Vorlesung 1902/03*. Dordrecht: Kluwer (2001).
Husserl, E. (1902/03b). *Logik, Vorlesung 1902/03*. Dordrecht: Kluwer (2001).
Husserl, E. (1905). Husserl an Brentano, 27. III. 1905. In his *Briefwechsel, Die Brentanoschule I*. Dordrecht: Kluwer (1994).
Husserl, E. (1906/07a). *Einleitung in die Logik und Erkenntnistheorie, Vorlesungen 1906/07*. Dordrecht: Martinus Nijhoff (1984).
Husserl, E. (1906/07b). *Introduction to logic and theory of knowledge*. Dordrecht: Springer, 2008. My translation of Husserl (1906/07a).
Husserl, E. (1908/09). *Alte und neue Logik, Vorlesung 1908/09*. Dordrecht: Kluwer (2003).
Husserl, E. (1913). *Introduction to the logical investigations, a draft of a preface to the logical investigations*. The Hague: M. Nijhoff (1975).
Husserl, E. (1917/18). *Logik und allgemeine Wissenschaftstheorie, Vorlesungen 1917/18, mit ergänzenden Texten aus der ersten Fassung 1910/11*. Dordrecht: Kluwer, 1996. My English translation, *Logic and general theory of science*, is forthcoming from Springer Verlag (2019).
Husserl, E. (1929). *Formal and transcendental logic*. The Hague: M. Nijhoff (1969).
Husserl, E. (1939). *Experience and judgment*. London: Routledge and Kegan Paul (1973).
Husserl, E. (1994). *Early writings in the philosophy of logic and mathematics*. Dordrecht: Kluwer.
Nelson, L. (1916). Extracts from Leonard Nelson's letter to David Hilbert. In Hill & da Silva 2013, 388–391.
van Atten, M. (2010). Construction and constitution in mathematics. *The New Yearbook for Phenomenology and Phenomenological Philosophy, 10*, 43–90.

Part III
Phenomenology and Constructivism as a Dialectical Relation

Chapter 7
The Truth of Proof: A Hegelian Perspective on Constructivism

Vojtěch Kolman

Abstract In the constructivist movement, which started with Brouwer's attack on the principles of classical logic and mathematics, the history of mathematics seems to repeat itself in a specific, self-conscious way. The main objective of this paper is to describe this approach by adopting the phenomenological method of Hegel. Its starting point consists in looking at knowledge as a continuous, yet painful, process—the Calvary of the Spirit—with its stations corresponding to the naive, direct concepts of experience as based on sense certainty or belief in the independent realms of objects and their transformation into more sophisticated, socially charged theories. The signs of this advance are the patterns of self-consciousness, such as Cantor's diagonal results, adopted by constructivism in a different but still powerful way. The key concept against which the progress of this development will be measured within the constructivist movement is the concept of proof, particularly with respect to Gödel theorems and the resulting split of knowledge into proof and truth. I will read this split, with the help of Lorenzen and Brandom, as a relative differentiation between two co-dependent aspects of self-consciousness that are to be prospectively conceived as two idealized dialogue partners.

Keywords Intuition · Brouwer · Diagonalization · Truth · Proof · Constructivism · Self-consciousness · Hilbert · Hegel · Lorenzen · Gödel theorems

7.1 Introduction

The basic idea behind my paper is to look at the constructivist attempt to analyze the foundations of mathematics as a *historical*—rather than purely theoretical—phenomenon, in which the key concepts such as *intuition*, *construction*, *truth* or *proof* play a dialectical role by being dependent both on the previous development and the competing foundational movements as well. I call this perspective Hegelian

V. Kolman (✉)
Institute of Philosophy, Czech Academy of Sciences, Prague, Czech Republic
e-mail: vojtech.kolman@ff.cuni.cz

C. Weiss (ed.), *Constructive Semantics*, Logic, Epistemology, and the Unity of Science 44, https://doi.org/10.1007/978-3-030-21313-8_7

by treating the given concepts "phenomenologically" in Hegel's specific sense of the word. In principle, this perspective is based,

First, on the difference of what things (mathematical objects, truths, etc.), are *in itself* and what they are *for us*,

Second, on the observation that this difference is *relative* and

Third that this relativity is responsible for the permanent fluidity and *historicity* of all concepts and truths, including those of mathematics.

I elaborated on these points to a great extent, though implicitly, in my book *Zahlen* (Kolman 2016a) and to a small extent, but explicitly, in the paper "Intuition and the end of all-isms" presented at the workshop "Constructive Semantics" (Friedrichshafen, September 2016). Some parts of this paper have been presented on different occasions, particularly in Kolman (2009, 2016b).

In the following text, I have chosen a bit of a different route, this one focusing on the concept of *proof* as being narrowly connected to constructive semantics and constructivism in general, particularly in its relation to the concept of *truth*, to which the very word "semantics" points to. Drawing on the phenomenon of Gödel theorems, I claim that the proof itself, when one-sidedly used as an exclusive point of departure, transforms itself immanently into a much broader notion which both cancels and retains its former form and contributes to the development of the respective mathematical experience. In the given case of Gödel's arithmetical sentence that is *proved* to be *unprovable*, and because of that to be *true*, one must simply differentiate between proof in two different senses of the word, as pointed out already by Zermelo in his unjustly infamous correspondence with Gödel (2003, p. 431).

Today, this particular finding has been transformed into the seemingly necessary and *absolute* difference between the purely formal, uninterpreted theory and its model, against which the truth of the theory's theorems can be measured. In my paper, I argue that what we have here is, in fact, the *relative* difference between the theory *for us* and the theory *in itself* which, however, is and must be in itself only *for us* or, as Hegel calls it, *in-and-for-itself*. For understanding these differences as constituting the progress of knowledge, Hegel's *Phenomenology of Spirit* is a valuable source of inspiration. Its interpretative benefits consist not only in its relation to the currently established narratives, such as that about the inside and outside of some theory, as known, e.g., from standard readings of the Löwenheim-Skolem paradox, but mainly in showing where these differences come from and how they relate to each other.

My exegetical line will be given by what Hegel called the "Calvary" of the absolute Spirit (Hegel 1977, § 808), with its stations corresponding to (1) sense certainty, (2) perception, (3) understanding, (4) self-consciousness, (5) reason and (6) the (absolute) spirit in which the journey shall, prospectively, complete itself.

7.2 Either-Or

7.2.1 Eleatics

When speaking of constructivism in mathematics one usually means a 20th century movement that basically started with the intuitionism of Brouwer and its attack on some principles of classical mathematics and logic—particularly the Law of the Excluded Middle. It then continues with the foundational theories and schools of Weyl, Hilbert, Lorenzen, Bishop, Markov or Goodstein.

As for the *Law of the Excluded Middle*, the cause of the constructivists' discomfort is relatively simple. Its traditional reading amounts to the possibility of arriving at the given conclusion *indirectly*, by way of the underlying *Either-Or* dichotomy. In my reading, this dichotomy (or difference, as I will call it) is something discovered by the Eleatic sophists who, in the work of Parmenides and his pupil Zeno, implemented the following influential method of scientific discovery:

(1) Somehow, by means of pure reasoning, I prove that hypothesis A leads to some unacceptable conclusion ("contradiction").
(2) From this, I infer without further ado that the negation of A must hold in the world, no matter whether I have some direct evidence of that or not.

So, e.g., I can prove, as the Eleatics did, that there is no change in the world by presupposing the hypothesis that object a is changing. Now, if a is changing then it means that it is not, in the end, a, i.e. a is not a, which is a contradiction from which, according to the Eleatics, the immovability of a and, derivatively, of the whole world follows.

The clear incongruity of this conclusion with our experience is "solved" by the *ontologization* of the Either-Or dichotomy which consists in splitting our world in two: (1) the ideal, motionless one, in which the true knowledge (*episteme*) obtains, and (2) the evanescent world of our senses, which is the world of a mere opinion (*doxa*). The difference between the truth *in itself* and the truth for us is slowly arising here.

7.2.2 Plato

Plato's theory of ideas not only further elaborates on this distinction, but also deals with the problem of the given split as based on a simple transition from A to its straightforward negation, or from the world of our senses ("Here and Now") to the world of abstract ideas (its "Beyond"). This is what the talk about *participation* (*methexis*) of the former in the latter is about and what is shown as questionable via the Third Man Argument (see Plato 1961, 132a-b; Aristotle 1998, 990b17, 1039a2, 1059b8, 1079a13).

Plato's style of writing dialogues in the *aporetic* form might be seen as a way of putting the Eleatic method to the test: One is not generally justified to infer from the fact that supposition *A* leads to the contradiction that non-*A* is valid, if only because non-*A* can lead to such contradiction too. This is, in fact, what Socrates usually brings his dialogue partners into. Obviously, he does not do it to doubt or reject the Either-Or dichotomy as such, but to show that its simple *formal* reading, as introduced by the Eleatics, is too weak to be held as valid in the general sense (see Stekeler-Weithofer 1995, Chap. 2, for a further discussion of this point).

In this general sense, the parallel derivability of the contradiction from *A* and from non-*A* does not have to be a case of mistaken argumentation, but a simple warning that the dialogue partners do not yet know what they are talking about. The later version of the Third Man Argument, the paradox of Russell, with the contradiction being deduced from the supposition that the set of all sets that do not belong to itself belongs to (or participates at) itself and *vice versa*, shows that the concept of set is unclear and worthy of further elaboration. As such, Either-Or might be read as an impulse towards conceptual clarification: The concepts shall be specified in such a way that one knows what is the case for *A* to be true *and* what is the case for *A* to be false.

7.2.3 Kant

Constructivism might be seen as a specific attempt at such clarification. Generally, it goes back to Kant's transcendental philosophy that stresses the fact that all truths and objects one can know are conditioned by our cognitive faculties and thus essentially *for us*. Hence, the mathematical truths cannot just be proclaimed true or false on the basis of some intellectual insight, such as indirect proof, but must be construed as true, or proven, in the present intuition.

However, despite this underlying constructive tendency, Kant not only did not cancel or clarify the dichotomy of Either-Or, but even multiplied it: first, by introducing it on the subjective ("for us") level when differentiating between intuition and concept as products of the faculties of sensuality and understanding and, second, by stipulating the thing "in itself" as the inaccessible, vanishing point of our knowledge. Hegel famously criticized the second point as illusory, based on what he called "bad" infinity and derivatively "bad" negation. This consists in the formal—and thus "bad"—reading of the Either-Or dichotomy.

7.3 Bad Negation

7.3.1 *Infinitesimal Quanta*

Hegel discusses the case of bad negation in connection with Newton's introduction of the *infinitesimal quanta* (see Hegel 2010, pp. 217–227.). Newton introduces them in the Eleatic style, i.e. by way of simple denial of some difference. As already stated, this cannot lead, at least not immediately, to a new kind of quantity but only to a state of conceptual disorder, as has been manifested in the history of mathematical analysis. Hegel's point is quite general here, applying to the previously discussed cases as well. Plato's theory of ideas, e.g., the observation that things around us are imperfect in many ways, leads, by way of the simple denial of this imperfection, to the idea of a *perfect* world in which these imperfections do not happen. As a result, though, *we* do not happen in this world either.

In the *Phenomenology of Spirit*, this state is described as the unhappy consciousness that is well-aware of the split between the world "here and now" and its "Beyond", yet does not know how to bridge it (see Hegel 1977, § 207–230). Hegel's solution to this situation consists in the realization that this Beyond is not something beyond ourselves, transcending us completely, but Beyond within ourselves, and generally, that the difference between things for our consciousness and the things in itself is made within consciousness itself.

As for the infinite quanta, this leads to their elimination in favor of the approximative processes in a particular dialectical way: These processes are not conceived of as determining some quantity a by another quantity i called infinitesimal, but as the very quantity a they are about to approximate if fulfilling some specific condition. So, two processes of determining some number (such as 1, 1/2, 1/4, 1/8, … and −1, −1/2, −1/4, −1/8, …) are considered to represent the same number (0) if and only if they fulfill Cauchy's condition of convergence both individually and with respect to their partial differences. The critical cases are those such as 3, 3.1, 3.14, 3.142, …, in which the approximated object (π) does not exist yet within the original realm of (rational) numbers and *must* be identified with this sequence itself. Put in Hegel's terms, its *for itself* is turned into its *in itself*, thus becoming the *in-and-for-itself*.

7.3.2 *Irrational Ratios*

A more general and deeper wisdom concerning the nature of knowledge is contained in this insight into the constitution of numbers. Namely, that these sequences do not describe nor approximate some purely external object of knowledge—because such an enterprise would be condemned to a failure from the very beginning—, but somehow produce or construct this object in an immanent way. Hegel's cryptic words from the introduction to his *Phenomenology* (Hegel 1977, § 85), that with the change of our knowledge its object changes too, must be read as an analysis of this situation in which one does not have any absolute ground against which the progress and validity of knowledge might be measured: There are only relative measures of the so far accepted theories that have proven themselves.

Hence, e.g., the proposition that all proportions are commensurable *in itself*—which the old Pythagorean mathematics took for granted and used in further demonstrations—turned out to be only true *for us* when the incommensurable magnitudes have been discovered. This did not lead, immediately, to the emergence of new—irrational—quantities, but only to a situation where such an emergence became possible, arising from the transformation of the conceptual incongruity (the given proportions, e.g., that of the diagonal and side of a square, are existing, yet rationally inconceivable) into the conceptual necessity, which took more than eighteen centuries to become true. The modern concept of the real *number* was made possible only due to Descartes' decision to identify one line segment as the *unit* E and to reduce the proportions of magnitudes A, B to the magnitude C such that $A{:}B = C{:}E$. The modern definitions of reals such as Cantor's and Dedekind's follow. (For a more detailed development of the concept of real number, see my book Kolman 2016a.) The given lesson is that yesterday's "in itself" is today's "for itself" and that this is not a deficit but an inherent pattern of knowledge which, so to speak, must construct itself from its own resources.

7.3.3 *Aufhebung*

Hegel pointed out that in this development there is a phase based on a simple negation of the previous state in which nothing is settled yet even though it might look to be so. This corresponds to the formal reading of the Either-Or dichotomy.

Now, one can say that the process of knowledge consists both in the negation of its previous phases (such as in the case of the universal commensurability of magnitudes) *and* its transformation into a new phase in which the old phase is not only negated, but retained as well by being duly specified (the introduction of *alogoi logoi*). This is what Hegel calls "Aufhebung" (see, particularly, Hegel 2010, pp. 81–82) and what, in fact, might be seen as related to the constructivist approach and its critique of the formalists' reading of logical and mathematical laws. Put differently, the constructivist reading of Either-Or does not allow one to proceed directly from the exclusion of the Either to the confirmation of the Or because there is typically a lot of work still to be done. The world is not simply there to be described and measured, if only because our descriptions and measures are a substantial part of it, thus contributing to its substantial change.

7.4 Sense Certainty

7.4.1 Forms of Intuition

In the rest of my paper, I will roughly follow Hegel's plan of the development of consciousness and his "stations of the cross" as labels for the reconstruction of phases that the constructive mathematics went through. Besides the work of Kronecker and other pre-constructivists, mathematical constructivism *stricto sensu* starts with Brouwer's (1907) decision to base mathematical truth on the intuition of time as a continuation of Kant's concept of "Anschauung". "Intuitive" and "intuition" are taken here in the first of their two extreme readings, namely as something evidently and unobjectionably true as opposed to the common use of something given without rational support and, as such, open to doubts and prospectively fallible.

Now, according to Brouwer, Kantian "intuition"—introduced also in the controversial form of "pure" intuition—turned out to be of the second, unreliable kind as far as its role in geometry is concerned, which is easily demonstrable by Kant's (1781/1787, A716/B744) own example of an intuitive truth, the geometrical theorem that the sum of all angles of a triangle equals two right angles. Kant borrows his intuitive proof of this theorem from Euclid's *Elements*, including its dependency on the *Fifth Postulate* which, as Riemann and Lobachevsky have shown, is independent of other axioms. This throws serious doubt on the apodicticity of the given reasoning, only to be amplified by the existence of "intuitive" models such as those of spherical geometry in which Euclidian geometrical space is conceived only as a local approximation of the global Non-Euclidian space. What these models and the associated reasoning show is that the intuition of space is dependent and mediated by theoretical stipulations such as Euclid's Fifth Postulate and, as such, cannot be "pure" in the Kantian sense of the word. The question now is whether Brouwer's intuition of time, whatever it is, is immune to this objection.

7.4.2 Free Choice Sequences

Knowing the fate of Brouwer's intuitionism and its eccentric theorems—such as that every total function on real numbers is continuous—the answer seems to be easy. But let us be more analytic and thorough about that. What one starts with then is a supposition that mathematical knowledge must be based on the most immediate and certain grounds which, according to Brouwer, is the intuitive experience of the *creating subject*. (See, particularly, Brouwer 1975, pp. 72–101 and 522–524.) Following the route of Hegel's *Phenomenology of Spirit*, this corresponds to the state of consciousness known as "sense certainty" that considers the knowledge given in the particular "here and now" as the most certain, rich and tangible, whereas for Brouwer this "here and now" limits itself to the inner experience of the creating subject's mind.

The fate of this phase of consciousness is also well-known: By not being able to differentiate itself from other episodes of knowledge such as "there and then", its truth turns out to be the most general and abstract that there prospectively is, thus transforming itself into another phase of knowledge called perception in which the impressions of the subject's mind are replaced by empirically demonstrable objects. Brouwerian variations on this theme might be found in the concept of the *free choice sequence* on which his theory of real numbers is based.

The point of this concept is that sequences by means of which numbers are identified are, in general, not given by some static law—capturing them as they are *in itself*—but freely and substantially developed and developable in time. And this is where the specific immediacy comes from: One does not work with the given sequence as a whole, but operates with the fragment constructed so far. This, of course, significantly affects the resulting theorems. So, e.g., the uncountability of reals is proven as a direct consequence of the observation that no mapping of reals on natural numbers—as based only on the initial fragment of the given real number—can be one-to-one because two different real numbers with the same initial fragment must take us to the same natural number. And, quite analogically, the theorem is proved that every total function on a continuum must be continuous, which is obviously false from the classical point of view.

7.4.3 Solipsism

A disquieting feature of these results, as delivered and developed by Heyting (1956, pp. 32–48), consists exactly in deliberately ignoring the fact that there is more to a given sequence than the "here and now" of its initial fragment, as indicated in the simple question: the fragment of *what*? The Cartesian certainty for mathematical knowledge that Brouwer intends to give us is, in fact, only illusory, based on the impoverishing and fuzzifying of the underlying idea of what real numbers as objects are in itself by focusing on their subjective "for us" part. This results in solipsistic foundations of mathematics and clever, yet, idiosyncratic re-definitions of traditional concepts such as continuity, continuum or sequence that only a few can actually follow.

This is to say that no matter how clever and interesting Brouwer's adjustments of classical mathematical concepts and truths are, they are also quite *arbitrary* and this arbitrariness does not amount to a decision to start in this or the other way, but to a failure to transform this or that state as it is for us into a state in which it is for us in itself, i.e., independent of my or your subjective decision or state of knowledge.

This shows us the double-faced nature of Brouwer's attack on classical mathematics and logic, as is aptly expressed by Bishop and Bridges (1980, p. 9):

> Brouwer became involved in metaphysical speculation by his desire to improve the theory of the continuum. A bugaboo of both Brouwer and the logicians has been compulsive speculation about the nature of the continuum. [...] In Brower's case there seems to have been a nagging suspicion that unless he personally intervened to prevent it, the continuum would

> turn out to be discrete. He therefore introduced the method of free-choice sequences for constructing the continuum, as a consequence of which the continuum cannot be discrete because it is not well defined. This makes mathematics so bizarre it becomes unpalatable to mathematicians, and foredooms the whole of Brouwer's program. This is a pity, because Brouwer had a remarkable insight into the defects of classical mathematics, and he made a heroic attempt to set things right.

To sum up: The Eleatic introduction of the Either-Or dichotomy, that Brouwer's foundational insights were aimed at, suffered heavily from being formal. At the same time, however, it was an attempt to dispose of the given arbitrariness by stipulating the objective Or as the etalon against which the right or wrong of the subjective experience might be measured. The question is whether one can have both: the objectivity of knowledge as based in the Either-Or difference without its purely formal reading as stemming from the vehicle of bad negation.

7.5 Perception

7.5.1 Infinity

Hilbert's famous reaction to Brouwer's subjectivist account of arithmetic can be seen as the first in a line of attempts at its rectification corresponding to the perception chapter in Hegel's *Phenomenology of Spirit*. As such, it follows two interconnected goals, namely to ensure the maximal *certainty* for mathematical experience while keeping it *objective* by transposing Brouwerian mentalistic construction from the inner form of the intuition to the public space of empirically controllable symbolic objects.

Hilbert arrived at this standpoint not only with respect to the "Bolshevism" of Brouwer, but also in response to the infinite quality of (mathematical) knowledge dealing with the seemingly infinite—and thus perceptually indemonstrable—objects. See particularly Hilbert (1926). Hilbert's solution corresponds to what Wittgenstein (1964, § 135) phrased as follows:

> 'We only know the infinite by description.' Well then, there's just the description and nothing else.

Hence, in mathematics or in other fields of experience we are dealing neither with some abstract and possibly infinite objects, such as infinite sets of set theory, nor with their subjective counterparts, but only with their finite descriptions, i.e. symbols representing them, either directly or by means of sentences referring to them. Our experience thus remains within the limits of the given empirical intuition that is supposedly both objective and certain.

7.5.2 Finite Einstellung

The account of mathematics that consists in dealing only with symbols as opposed to abstract entities is known as formalism and was—already before Hilbert's times—attacked by Frege (1893/1903, § 93). He identified two critical problems:

First, there is a conceptual difference between what some representation is and what it represents, even in the abstract cases of numbers, because the numbers and not the numerals or mental images have objective properties such as being prime or even, as opposed to the properties such as fuzziness, to be written in ink, Arabic, etc.

Second, mathematics defined in a formalist way as a simple symbolic game according to some rules does not allow us to directly introduce the concept of truth: The rule that a knight in chess might move in this or that way is neither true nor false but is simply so. What might be true or false are the meta-sentences about the possibilities of achieving this or that goal with this or that rule, e.g., to checkmate the opponent's king in two moves.

Hilbert (1926, p. 161, 1930, p. 385) addresses both these objections by noticing that it is not the symbol itself that is under scrutiny in formalized mathematics but its *use* within what he later called the "finite Einstellung". According to his own words, it is to be thought of as a new kind of Kantian intuition consisting of working with symbols "here and now", but this time not privately, but within the outer spaces of sense perception. The discipline of so-called metamathematics, accordingly, does not deal only with symbols but also with their derivability by means of certain calculi.

7.5.3 Fallibility of Perception

The resulting concept of knowledge, though, turned out to be very far from the original goals. To demonstrate that some series of symbols is derivable is to give a specific derivation, i.e. another series of symbols, of which I must be certain that it is the right answer to the original question. This certainty, though, transcends the mere existence here and now, which is particularly clear in the case of the conversion that something is *not* derivable.

Here, by definition, the respective proof must go beyond the problem of the empirical demonstrability and thus be mediated by rational arguments which, again, might be right or wrong. This makes the whole enterprise susceptible to a higher-order mistake, as witnessed by Bernays laconic commentary to Hilbert's (1935, p. 210) collective works:

> it has turned out that in the realm of meta-mathematical reasoning the possibility of a mistake is particularly great.

In the light of this, one can say that this is a sign that intuition—be it in the inner, apodictic form of Brouwer or in the outer, empirical form of Hilbert—does not play any role in mathematics. And this is, in fact, what Frege and other of the so-called logicists inferred long before Brouwer and Hilbert came on stage, claiming that mathematics is of a purely conceptual—or logical—origin. But this is different story, and one that I have discussed in detail elsewhere, including the relation of Frege's logicism to Brouwer's and Hilbert's programs (see, e.g., Kolman 2015). Here, the lesson to be learned is a rather different one: namely, that if intuition should play any role in mathematics at all, it cannot be infallible, i.e. certain and objective at the same time.

7.6 Understanding

7.6.1 Logical Machine

The fallibility of intuition and perception seems to call for a more radical objectivist articulation of knowledge than its mere placement in the empirical space governed by the "finite Einstellung" in which it is still dependent on the observing subject. This leads to the idea of the intuitiveness and certainty of mathematical results being guaranteed by their mechanical checkability by means of a suitable device—i.e. an independent *object* known since Lully's times as *a logical machine*. Such a machine or computer, though, is always an instantiation of the theoretical concept—today known as the so-called *Turing machine*—, being the sum of some abstract, yet effectively feasible rules according to which some symbols might be used. This yields the concept of knowledge based on abstract rules and laws standing behind the empirical reality of our concrete use, and, as such, it roughly corresponds to the form of knowledge introduced in the *Phenomenology of Spirit* as understanding, in the sense of Kant's original concept "Verstand", the faculty of rules.

Historically, the idea of mathematical knowledge being guaranteed by the Turing machine did not come from Hilbert, but from Gödel—though not under this name, but in a form that was proved to be equivalent later. He didn't need it to reinforce finitist methods but to make them more exact and explicit in order to show their limits. The meaning of Gödel theorems, accordingly, can be explained as follows: There is no way of rooting the objectivity of mathematics in the primitive forms of calculation guaranteed by the logical machine since such an objectivity will always be overcome by its broader version as stemming from the subject that invented and must control the machine.

Of course, this looks like a return to the original, subjective form of knowledge; in fact, it deliberately is so but not in the original version of Brouwer in which one simply discarded the objective Or in the Eleatic Either-Or distinction. I will arrive at this reading later, within the context of Gödel's result. It should be obvious already now, though, that this reading is of a social nature, based on the insight that the calculations given by the Turing machine can be independent of us only to the extent that *we* see them as right or wrong on their own, where this *we* now adopts the role of the Or to the Turing machine's Either.

This exact point was made by Wittgenstein (1953, § 193) with respect to the related problem of rule-following: When using a machine to symbolize a particular action of this machine, says Wittgenstein, we often forget the possibility of its parts' bending, breaking off, melting and so on. Thus, in the end, it is we again who must guarantee whether the machine works according to our plan and who recognizes itself in the abstract rules behind the subjective reality of our sensuous, empirical life. In the *Phenomenology of Spirit*, this is marked by a gradual transformation of the chapter dealing with understanding into one dealing with self-consciousness.

7.6.2 Peano Arithmetic

In Hilbert's project, this turn had been prepared by symbolic finitism gradually adopting the specific form of so-called *axiomatism*, in which the symbols that one is dealing with in the present intuition are not quite arbitrary, but are representing *formulas*. This seems, at least partially, to take into account Frege's objection that it is not derivability itself which we are interested in our reasoning, but the arithmetical truth pertaining to whole sentences as opposed to their non-sentential parts. See Sect. 7.5.2.

At the same time, the traditional interpretation of axioms as being evidently and irrefutably true was replaced by their conventionalist reading in which the semantic role of truth was assumed by the syntactic concept of consistency, i.e. non-derivability of some contradictory sentence (such as $1 \neq 1$) from the axiomatic whole. In the consistent axiomatic system, first, every derivable formula is to be proclaimed true just on the basis of the consistency of the given axioms and, second, derivations are both produced and checked by a Turing machine. The axiomatic system itself is conceived as a mechanically describable set of formulas and rules which are to generate other formulas from them.

The exemplary case fulfilling these conditions is the so-called Peano Arithmetic (PA) given by the following axioms:

$$\begin{array}{llll} \text{(P1)} & \forall x \neg(x+1=0), & \text{(P2)} & \forall x \forall y (x+1=y+1 \rightarrow x=y), \\ \text{(P3)} & \forall x (x+0=x), & \text{(P4)} & \forall x \forall y (x+(y+1)=(x+y)+1), \\ \text{(P5)} & \forall x (x \times 0=0), & \text{(P6)} & \forall x \forall y (x \times (y+1)=(x \times y)+x), \end{array}$$

and the axiomatic scheme

$$\text{(PI)} \quad (A(0) \wedge \forall x (A(x) \rightarrow A(x+1))) \rightarrow \forall x A(x).$$

(PI) is known as *mathematical induction* and, as a scheme, it represents the infinity of axioms of the given form which, however, is not at variance with the formalist principles since these axioms are still mechanically recognizable as such by a Turing machine. The strictly finite part of (PA), which amounts more or less to the axioms (P1–6) without (PI), is known as the so-called Robinson Arithmetic (Q) and is sufficient for the demonstrability of the essential part of Gödel's incompleteness results.

The inferential rules connected to (PA) and (Q) are those of the underlying predicate calculus, consisting of several logical truths and principles such as modus ponens and generalization:

$$\text{(MP)} \quad A, A \to B \Rightarrow B, \text{ (G)} \quad A(x) \Rightarrow \forall x A(x).$$

7.6.3 *Reflexivity*

It is against this specific background that Gödel (1931), in his incompleteness results, comes with the following explanatory twist: He shows that to *every* formal axiomatization (T) of arithmetic strong enough to the extent that it "contains" at least the system (Q), there is a sentence which is not derivable in it but at the same time mathematically true. In fact, this sentence is true not *in spite* of its being underivable in the given formal system but exactly *because* it is underivable in it: it is formed so as to express the very idea that it cannot be derived in (T).

Here, at the same time, both the subjectivity of the given intuition and the objectivity of the calculating machine are broken through and lead to re-establishing the Either-Or difference by showing the insufficiency of the Either side to provide for the mathematical truth. Two important points are as follows:

First, the underlying concept of proof is adjusted by the concept of truth via the possibility of proving that something is improvable in the given axiomatic system, which obviously transcends the underlying concept of finite or mechanical provability.

Second, this transcendence is enabled by the possibility of arithmetical sentences to refer to themselves which explicitly marks the entry of *self-consciousness* into the mathematical experience.

One can describe the whole situation in Hegel's terms as follows: *The truth of the mechanical concept of proof turned out to be another, broader concept of proof which is to be, consequently, identified with the very notion of truth.* The concepts of proof and truth are the new horns of the old Either-Or dichotomy, this time arising within the subject itself by means of its ability to reflect on itself, i.e. to be self-conscious.

As it will turn out later, such a split within the subject itself corresponds, in fact, to the intersubjectivity of two subjects mutually interacting and checking on each other. In the following section devoted to Lorenzen's operative reading of Peano axioms, this move will be pre-established by sketching Lorenzen's later idea of dialogical logic in its relation to Hilbert's program and Gödel theorems. The full exploitation of this progress, quite in accord with Hegel's own progress in the *Phenomenology of Spirit*, will come into play later, under the title of spirit.

7.7 Self-consciousness

7.7.1 *Operative Mathematics*

Let us start with the revised version of Hilbert's concept of operating within the "finite Einstellung" transformed into the objective derivability by means of a Turing machine. The fact that some arithmetical sentence is true seems to be reducible to its derivability in the given formalism such as (PA), supposing it is consistent. But of course there are some previous expectations and requirements put on this truth as stemming from the fact that arithmetic is primarily based and used in practical calculations and is only derivatively connected to calculations in (PA).

In accordance with this, Lorenzen (1955) in his operative logic and arithmetic suggests seeing (PA) primarily not as a set of uninterpreted and quite arbitrarily chosen formulas but rather as a means of an advanced explicitation of the derivability in other calculi that are in their overall design quite specific for arithmetic. Using, for simplicity's sake, the unary notation, these might be, e.g.:

$$
\begin{array}{llllll}
(|) & \Rightarrow |, & (=) & \Rightarrow | = |, & (<) & \Rightarrow | < x| \\
 & x \Rightarrow x|, & & x = y \Rightarrow x| = y| & & x < y \Rightarrow x < y| \\
(+) & \Rightarrow x + | = x| & (\times) & \Rightarrow | \times x = x & & \\
 & x + y = z \Rightarrow x + y| = z| & & x \times y = p, p + y = q \Rightarrow x| \times y = q. & &
\end{array}
$$

They are introducing, respectively, the number series (|), equality (=), operations of addition (+), multiplication (×) and the relation of "less than" (<). The internal difference between these calculi consists in using the variable in two possible ways, namely as the so-called *Eigenvariable*, which generates its range, as in the case of (|), and the *object variable* which presupposes its range as already given, as in the rest of the calculi that simply refer to symbols produced by (|).

7.7.2 *Infinity of Meaning*

Now, to justify a sentence like this $||| < ||||$, one must start with the series $| < ||$, which is derivable according to the first rule of (<), and continue with $|| < |||$, $||| < ||||$ according to the second rule of (<). But to justify the generalized sentence such as the unary version of (P1)

$$(\text{P1}') \qquad \forall x \neg (x| = |),$$

such a simple justification is not available because, as we know, what one has to show is not the particular derivation, but the non-existence of *any*. This makes the task not only negative, but infinite as well.

This infinity of arithmetical meaning is, obviously, further developed by the general form of (P1′) as marked by the use of the quantifier. In the case of (P1′) this amounts to the claim that $\neg(x| = |)$ is justifiable for *any* numeral substituted for x, which consists in simultaneous derivability of all these infinitely many instances. And it is exactly this combination of negativity with infinity that makes Hilbert's finitism unsustainable: The presupposed finitude of the given symbols turns out to only be apparent, depending in its meaning on the infinity of the symbols' operative use. This use, though, is reducible neither to empirically observable behavior nor to the actual operations of the Turing machine, but to the broader concept of what might be called "operative Handlung"—"operative act" as something which the subject cannot do on its own but only with respect to other subjects.

This very insight took Lorenzen from his operative logic and mathematics to the broader concept of dialogically based logic, mimicking the original transformation of the Kantian sphere of constructions within the pure intuition into the Hegelian realm of socially articulated spirit. See Lorenzen and Lorenz (1978). This, of course, might be a bit of a surprise within the seemingly de-socialized context of mathematics, but the basic idea is simple enough to be generally acknowledged as true: To show that the figure a is *not* derivable, one can either offer some rational argument why that is, or, as Lorenzen suggests, ask somebody to contradict it by claiming the opposite—the derivability of a. As such, it is still the derivability, i.e. the existence of some derivation which is to be shown in the first place but in a different situation of a dialogue between two people. One of them is a proponent of the claim that a is not derivable, the other an opponent, claiming that a is derivable.

7.7.3 Dialogical Logic

Accordingly, one might interpret the discussion about infinity as follows: It does not stand primarily for the fact that in mathematics, as opposed to other realms of experience, one deals with the infinite but instead for the fact that we are *always* interested in justifying the given sentence with respect to an *arbitrary* opponent by means of some *general winning strategy*. Only then are we granted the right to call the given sentence true. Now, this observation becomes the explicit part of the game for the case of the general quantifier. In the case of (P1′), for example, the winning strategy consists in the proponent's defence of $\neg(x| = |)$ against the arbitrary choice of the numeral produced by (|) for x. This is to say that the given opponent chooses some particular x, e.g., |||, and the proponent of (P1′) must commit himself to the particularized claim of $\neg(|||| = |)$. This might be attacked by committing to the negated thesis $|||| = |$, which, as follows from an easy, yet negative analysis of (=), *cannot* be derived. Obviously, the argument is valid independently of the choice of x and, as such, can be won against an arbitrary opponent.

Such a reconstruction of arithmetical semantics elaborates on the insight that truth is not the static property of some sentence but rather a dialogical process of its justification. From the logical point of view, this process is governed by the rules specifying what attack and defence rules are to be used when a proponent's claim has the form of negation, conjunction, disjunction, implication, equivalence or quantified sentence. Generally, there are also rules specifying when the dialogue is won (or lost) and what it means for the proponent to have some general winning strategy against an arbitrary opponent. Different fields of knowledge, such as mathematics or ethics, can have additional rules by which they operate.

In toto, the problem of the epistemic standard which makes the resulting concept of knowledge both objective *and* accessible to subjective needs seems to be solved by attributing it to another subject. "Infinity" might serve in this as a different name for this social stratification of meaning, or, put otherwise: *The truth of the operative treatment of symbols turns out to be the infinity of the social interaction.* The nature of such infinity is further elaborated on within Gödel's results concerning the incompleteness of the mechanically described calculi such as (PA) or (Q).

7.8 Reason

7.8.1 The ω-Rule

In order to understand the role of Gödel theorems in the development of the mathematical spirit, one needs to attend to the following: There is a significant difference between justifying the truth of some formula by means of (PA) and by means of the original arithmetical calculi ($|$), ($+$), ..., as in the case of ($P1'$) given above. In the latter case, the justification was based on a general strategy for any particular numeral—i.e. product of the calculus ($|$)—to which the object variable referred to, as opposed to the variables in (PA) which are left open to different interpretations and are thus independent of them. As a result, in order to justify a generalized sentence such as ($P1'$), one needs to justify the infinitely many instances of $\neg(x| = |)$, i.e. $\neg(|| = |), \neg(||| = |), \neg(|||| = |), \ldots$, though by means of a finite winning strategy. Contrary to this, in (PA) the derivability of $\forall x A(x)$ requires only the derivability of $A(x)$, as given by the rule (G).

Now, according to Gödel theorems, to every mechanically describable (or "recursive") axiomatization of arithmetic (T) such as (PA) that is powerful enough, i.e. contains at least (Q), there is an arithmetical sentence which is true but underivable. In the light of the previous differentiation, this is to say that there is some general winning strategy of how to justify this sentence with respect to the calculi ($|$), ($+$), ($\times$), ... but not with respect to (T). Since the given sentence is of the form $\forall x A(x)$, one can look at the difference as consisting in the employment of the so-called (ω) rule:

$$(\omega) \quad A(|), A(||), A(|||), \ldots \Rightarrow \forall x A(x),$$

i.e. the rule with infinitely many premises, as opposed to the finite (G). This leads to the specific concept of ω-incompleteness:

there is a sentence $\forall x A(x)$ for which every instance $A(|)$, $A(||)$, $A(|||)$, ... is deducible in (T), and as such true, but not $\forall x A(x)$ itself.

The additional fact that A is decidable shows that one can have a particular winning strategy for any $A(|)$, $A(||)$, $A(|||)$, ..., without having a general strategy for $A(x)$ as such. This situation is known from the so-called Goldbach Conjecture which shares with Gödel's critical sentence the form of the quantified sentence $\forall x A(x)$ with a mechanically decidable basis $A(x)$. In Goldbach's case, this basis says that any even number greater than 2 is the sum of two primes.

7.8.2 Full- and Semi-formalism

The important difference between the Goldbach Conjecture and Gödel's critical sentence might be phrased as follows: Though we have a winning strategy for every instance of $A(x)$ in both cases, only in the latter case do we know positively that all the instances are to be won and, as such, true. In Goldbach's case, we know only that we can decide whether $A(n)$ is true or false for any given n, because there is a finite way of verifying that n is an even number and that, if n is an even number, it is the sum of two primes. We do not know yet, however, that this will turn out to be true in every case. In Gödel's case, on the contrary, we know positively that every instance of $A(x)$ is to be won and, in this sense, we have a general winning strategy at our disposal. But this strategy, unlike the particular strategies for $A(|)$, $A(||)$, $A(|||)$, ... is not based on the calculations in (T) but on their extension by means such as (ω).

Going back to Schütte (1960), Lorenzen (1962) articulates this difference with the help of the concepts of *full-* and *semi-formalism*, the latter one being defined by employing the infinite rules such as (ω). In the end, though, this difference does not lie in the infinity of the semi-formalism, but in the specific way it will be described. These descriptions will be, as mentioned above, always finite, but in the case of full-formalisms such as (Q) or (PA) be based on more mechanical, schematic means, as captured in the specific concept of the Turing machine. The semi-formal systems might use more generous, though still constructive, interpretations that allow us to show, as in the case of Gödel theorems, that some general sentence is true in a way which transcends the possibilities of any full-formal system containing (Q). This transcendence is deliberately based both on the schematic nature of full-formal systems and the self-referential possibilities of arithmetical language.

The details of Gödel's proof are as follows. The arithmetical sentence is built in a way which says that there is no number coding the proof of this very sentence in the full-formal system (T) containing (Q), symbolically

(G) $\forall x(\neg \text{Proof}(x, g))$.

Here, g is the code of (G) and Proof(x, y) is some formula in arithmetical language capturing adequately the property of being a proof in (T) in the sense that two numbers m, n stand in this relation if and only if m codes the derivation of the formula coded by the number n. For brevity's sake, I will not symbolically differentiate between the number (code) and its symbolic representation (numeral), as required in the fully developed presentation of Gödel's incompleteness results.

Now, one can show that the formulas capturing the basic syntactic relations such as "to be an arithmetical sentence", "to be an axiom", "to be a proof", etc., are derivable in (T) if and only if they are justifiable by dialogical means, i.e. in arithmetical semi-formalism. This might also be phrased that they are derivable in (T) "in accord with the truth" if we are well aware of the fact that the concept of truth is what is under scrutiny here. In addition to this, there is an important reflective result saying, in a nutshell, that there is a number g such that it codes the formula gained by the substitution of g for y in $\forall x(\neg\text{Proof}(x, y))$. This is the formula (G). Its existence follows from the possibility to define the operation $\text{subst}(x, y)$ which yields for given numbers m, n the code q of the result of substituting n (or the corresponding numeral) for the sole free variable in the formula coded by m. Thus, the exact form of the formula (G) is, in fact, $\forall x(\neg\text{Proof}(x, \text{subst}(m, m))$, where m is the code of $\forall x(\neg\text{Proof}(x, \text{subst}(y, y))$ being the formula with y as a single free variable. Obviously, $\text{subst}(m, m)$ is code of $\forall x(\neg\text{Proof}(x, \text{subst}(m, m))$, the existence of which has been under the name of g stated in the reflective theorem mentioned above.

At this moment, the required result easily follows: Assuming that (G) is provable in (T) means that there is some code p of this proof, which leads to the derivability of $\text{Proof}(p, g)$ in (T). This is, though, inconsistent with the derivability of $\forall x(\neg\text{Proof}(x, g))$, from which the derivability of $\neg\text{Proof}(p, g)$ directly follows. Hence, supposing that (T) is consistent, the sentence (G) cannot be provable in (T), which means that every instance of $\neg\text{Proof}(x, g)$ is provable in (T) and, thus, (T) is ω-incomplete. But this means that the formula (G) is justifiable as true, though not by derivation within the full-formalism (T) but in the arithmetical semi-formalism in which the proponent of (G) has a general winning strategy by showing that the formula $\neg\text{Proof}(n, g))$ is provable in (T) for any n of the opponent's choice.

7.8.3 *The Truth of Proof*

What one has here is the progression from the mechanically defined concept of arithmetical proof, capturing the original idea of the purely symbolic foundations of mathematics, to the broader concept of arithmetical proof which somehow refers to its previous, narrower version and thus transcends it in an interesting and fruitful way. Traditionally, one reads this as a progression from the existence of the arithmetical proof to the necessity of arithmetical truth as captured in the modern model theory as opposed to the purely syntactical proof-theoretical considerations. But our decision to phrase this in the proof-theoretical language of full- and semi-formalism was led by the idea that this is, in fact, not a matter of two categorically different spheres—syntax and semantics, corresponding to the Either of the subjective mind and the Or of the objective world—but a homogenous split within mathematical self-consciousness. That is why this section corresponds to the reason (*Vernunft*) section of *Phenomenology of Spirit* with its main lesson being that our consciousness cannot be, for its own sake, detached from the external world but is essentially conditioned by it.

In light of this, one can read Gödel's reaction to Hilbert's symbolic program not as a mere rehabilitation of the original Either-Or split, but as its reconstruction in the deeply constructive manner in which the Gödelian semantic Or is not based merely on the negation of the Hilbertian syntactic Either, but on their constructive synthesis. This stems, as already said, from the fact that Gödel's sentence (G) is true not *despite* its unprovability in the given full-formal system, but exactly *because* of it. In this sense, the Either horn, the fact that something cannot be done by the given finite methods, is a part of the Or horn, the fact that it can be done in another way. Put in Hegel's terms: *The truth of the full-formal proof systems is the truth of their semi-formal extension, or more generally, it is our very capability of making these proof systems the object of our examination.*

7.9 Spirit

7.9.1 Diagonalization

The previous section, devoted to matters of reason, revealed that the possibility of arithmetical language to talk about itself, as in the case of (G), does not come from arithmetic itself, but from our possibility of reflecting on ourselves and our own doings, such as calculations according to some schematically described rules. The arithmetic here only represents the part of ourselves that is autonomous and rich enough to be able to manifest this fact in a relatively perspicuous manner.

Historically, such a possibility of transcending the given arithmetical methods by means of their reflection, as manifested in Gödel theorems, goes back to the diagonal method of Cantor and its use in "proving" the non-denumerability of reals as opposed to the denumerability of natural numbers. Here, again, the ontological reading of the result must be opposed to its cautious, constructive variant. What one has here, then, is not, as the standard reading suggests, the existence of a "bigger" set of numbers where some of them, because of the denumerability of language, must be unnameable and thus unrecognizable but only the possibility to add to some denumerable list of reals another one which is not on the list because it was built with reference to it. See my paper Kolman (2010) for further details.

In accord with this, both Cantor's diagonal arguments and Gödel theorems can be read in two basic ways roughly corresponding to the former two readings of the Either-Or difference. Gödel (1995, p. 310) himself described these readings as follows:

Either there exist *absolutely* unsolvable diophantine problems […], where the epithet absolutely means that they would be undecidable, not just within some particular axiomatic system, but by *any* mathematical proof the human mind can conceive.

Or the human mind (even within the realm of pure mathematics) infinitely surpasses the power of any finite machine.

The Either here refers to the formal reading of the Either-Or difference and the Or to its constructive version. The version which I have been reproducing in the previous sections acknowledges the necessity of the independent corrective to every positive concept of knowledge—be it sense certainty, use-theoretical approach or a mechanical concept of objectivity. At the same time, it does not place this corrective into the completely inaccessible *Beyond* defined in a purely negative way. Phrased in Hegel's terms, we are measuring the objective validity of a given theory, representing the theory for us as captured in the full-formal systems being under our exclusive control by means of the theory in itself, represented here by the more generously defined semi-formal systems.

7.9.2 Justified True Belief

The final step in the whole story of mathematical consciousness is provided by the reinterpretation of the given split between full- and semi-formalism along the lines suggested in Lorenzen's dialogical semantics. The idea behind this move consists in an insight into the social nature of our reason: The fact that I am able to think about myself, and even to think at all, stems from my ability to look at myself from the point of view of other people. Thus, it is other people who provide the objective Or to the subjective Either, thus preventing the subjective collapse of knowledge into a mere hallucination without leaving the realm of the human spirit.

Against the background of Lorenzen's dialogical foundations for arithmetical semantics, one can further elaborate on this overall concept in a way suggested by Brandom (1994). This starts with the traditional delimitation of knowledge as *justified true belief* (see Plato, *Theaetetus* 201 c–d) which, at first, seems to belong to somebody's private mind and thus leads to the concept of truth as this mind's episode or cognitive state. In the light of this, Brandom focuses on the concept of justification, particularly with respect to the so-called Gettier (1963) counterexamples. These are based on beliefs presented as true and justified but in a quite accidental, unconnected way. So, e.g., a farmer has some evidence in favor of the belief that his favorite cow is in a meadow—he saw, for example, the well-known brown and white "stains" through the bushes in front of him. And, in fact, the cow really is in that particular meadow, but not on the spot that the farmer believed he had seen it. The stains were an illusion created by the leaf. The immediate idea to rectify this by laying some further requirements on the concept of justification, such as a more direct, haptic contact with the cow, a genetical examination of it, etc., does not take us anywhere since, obviously, for every such rectification there always is and will be the theoretical possibility of doubting it simply because there is, by definition, the essential gap between the mind that adopts the given belief and the world to which this belief is related to.

What Brandom notices at this point is that in the whole story there is another person who—unlike the farmer—knows exactly how things really stand, namely the *narrator* of the whole story. It is he and not the farmer who says how things "really" are as opposed to how they only "seem to be", and, as such, he is also the only source of the given sentence's truth. In the light of this, Brandom's solution follows in the expected social way: Knowledge is not the state of a single mind but a complex social status consisting in the distribution of the normative attitudes between two idealized dialogue partners (see Brandom 2000, Chap. 3, for further details). The three steps contained in this are as follows: First, the basic form of knowledge is its ascription to somebody else in the form of a commitment, which is the belief part of the original definition. Second, the claim's justification lies in the ascription of the entitlement to the given belief, based, e.g., on the fact that the farmer is not colour-blind, does not have a history of lying, etc. And third, the truth part of the definition copies the narrator's role in the whole story, in which the ascriber takes the given belief and commitment to it for its own.

7.9.3 Absolute Spirit

In the case of Gödel theorems, we obviously have the ascriber's standpoint identified with the arithmetical semi-formalism as opposed to the arithmetical full-formalism to which some "belief", e.g., the derivability of (G), is ascribed. What happens now is the following turn of events: The sentence (G) is justifiable by means of the semi-formal reasoning which leads to the claim of the sentence's truth. But this truth is based on the impossibility to derive (G) by means of full-formalism alone. Hence, from the standpoint of full-formalism, one is not justified to believe in (G). What is important here is that, in this way, both standpoints are significantly changed: The ascriber adopts the fact of the inability of the ascribee to justify (G) as part of this sentence's truth, thus changing the full meaning of what he is ascribing.

What one gets is a stimulating and innovative example of what Hegel might mean by saying that with the change of knowledge its object is changing too. Obviously, the fact that the given object of knowledge is this knowledge itself plays a significant role in this. Accordingly, the difference between the theory for us and the theory in itself becomes obsolete, not in the sense that it does not play any role in the further development of spirit, but because it is itself part of what the theory—e.g., the arithmetical semi-formalism—takes into account, and as such is in-and-for itself. This provides the difference between the spirit and the previous stages of consciousness, such as sense-certainty or understanding, in which the truth of the theory, its "in itself", was not manageable by the theory's own means and thus different from what the theory was "for itself" and "for us". Hence, the very idea of the interpretative closure, as connected to Hegel's spurious concept of the *absolute*, can be read in this cautious way.

With respect to mathematics, I suggest modelling this difference on the *constructivism in the strong or dogmatic sense* of the word, which simply presupposes some methods of proof as definitive and given as opposed to what might be called the *dialectical constructivism* in which the immanent property of our theories to transcend themselves is somehow accounted for. I find this difference instantiated in Hilbert's early formalism or in the recursive analysis of Goodstein and Bishop on the one hand, and Lorenzen's operative and dialogical approach on the other hand. Gödel theorems represent here the testing case against which the expressive strength of the given theory might be measured. The promoted "absolute" quality of the latter approach does not consist so much in its perfect and irrevocable shape as in its essentially reflexive and self-conscious nature.

7.10 Conclusion

In my paper, I have followed Hegel's development of consciousness within the realm of mathematical spirit, showing how it proceeds from the primitive concepts of knowledge as based on the mathematical sense certainty and the simple objectivist reading of mathematical truths to the concepts rooted in social agency. I limited myself to the later phase of this development, as instantiated in the constructivist movement of the 20th century, but I easily might have started, as Kvasz (2008) did, with the ancient mathematics of intuitive demonstrations and proceeded, by way of their mediation with linguistic, e.g., algebraic means, to the birth of self-consciousness in the set theoretical frame of Cantor.

The chosen phenomenological method allowed us to see the concept of proof or intuition not as some fixed difference to be refused or rectified by the development of mathematical knowledge and the foundational concepts connected to it, but as a constitutive part of this development itself. In the case of intuition, one can thus proceed from its purely subjectivist and infallible meaning, as exemplified in Brouwer's reconstruction of Kant's original concept of pure intuition, through its empirical and objective reinterpretations in the work of Hilbert all the way to its broader social anchoring in geometrical practices such as those of forming solids and assessing the quality of their form. See Lorenzen (1984) and Stekeler-Weithofer (2008) for further details. In this way, one can even claim that Kant "was right" because, as Stekeler-Weithofer (Unpublished) put it:

> Pure intuition is a label for the mere form of [...] an objective reference to objects of perception—including the corresponding spatial and temporal transformations of perspectives if there are different observers at different places or if we refer to the same object or event from different times.

(See my paper Kolman 2018 for further details.) As for the concept of proof, which was the main subject of this paper, there is a similar and intersecting line of thought, enriched by the fact that it relates directly to judgements and to matters of truth. My reasoning was as follows:

Constructivism starts with the insight that the concept of truth—of what objects are in itself—cannot be completely detached from the concept of proving how they are for us. I discussed this in terms of the Either-Or dichotomy and its possible readings. As a result, my starting point was Brouwer's suggestion to identify the truth with the constructions provided by the creating subject. This yielded the cancellation of the Either-Or dichotomy and led to the problems described by Hegel in his chapter on sense certainty. From this, I proceeded to the problems of the momentaneous perception as given in Hilbert's symbolic foundations of arithmetic followed by the concept of understanding in which the real object of knowledge turned out to be the abstract rules hidden beyond the symbols and their use.

Here, as in the *Phenomenology of Spirit*, the purely subjective concept of knowledge transformed itself into concepts based on the subject-object distinction. Its unsustainability led to its further transformation, which made the object into the subject, i.e. the subject-object distinction into the subject-subject distinction, without falling into the trap of solipsism by reading this turn in a social way. This reading was elaborated on within the transition from self-consciousness through reason to the absolute spirit. In this, Gödel theorems served me not only as object but also as the vehicle of the whole change. First, there was the Hilbertian idea of a purely syntactically defined concept of arithmetical truth based on its reduction to the mechanically described operations with symbols. This idea was overcome by the possibility of arithmetic talking about itself, which, as it turned out, stems from our self-conscious nature. Thus, the truth of the proof, speaking Hegelianese, turned out to be its limited nature which led to the broader concept of proof to be traditionally identified with the concept of truth. In other words, the truth of the proof is the truth again, which might serve as a Hegelian gnomon of what "absolute" knowledge is about.

Acknowledgements The author would like to thank Professor Pirmin Stekeler-Weithofer for valuable conversations on the subject and Dr. Tereza Matějčková for the detailed comments on the final draft of the paper.

References

Aristotle. (1998). Metaphysics. (H. Lawson-Tancred, Trans). London: Penguin Classics.

Bishop, E., & Bridges, D. (1980). *Constructive analysis*. Berlin: Springer.

Brandom, R. (1994). *Making it explicit: Reasoning, representing, and discursive commitment*. Cambridge, MA: Harvard University Press.

Brandom, R. (2000). *Articulating reasons. An introduction to inferentialism*. Cambridge, MA: Harvard University Press.

Brouwer, L. E. J. (1907). *Over de grondslagen der wiskunde* (Ph.D. Thesis). Amsterdam: Universiteit van Amsterdam.

Brouwer, L. E. J. (1975). *Collected Works I*. In A. Heyting (Ed.). Amsterdam: North-Holland.

Frege, G. (1893/1903). *Grundgesetze der Arithmetik I–II: Begriffsschriftlich abgeleitet*. Jena: H. Pohle.

Gettier, E. (1963). Is justified true belief knowledge? *Analysis, 23*, 121–123.

Gödel, K. (1931). Über formal unentscheidbare Sätze der 'Principia Mathematica' und verwandter Systeme I. *Monatshefte für Mathematik und Physik, 37*, 349–360.

Gödel, K. (1995). Some basic theorems on the foundations of mathematics and their implications. In S. Feferman, J. W. Dawson, W. Goldfarb, C. Parsons, & R. M. Solovay (Eds.), *Collected works III* (pp. 304–323). Oxford: Oxford University Press.
Gödel, K. (2003). *Collected works V. Correspondence H-Z*. In S. Feferman, J. W. Dawson, W. Goldfarb, C. Parsons, & W. Sieg (Eds.), Oxford: Oxford University Press.
Hegel, G. W. F. (1977). *Phenomenology of spirit* (A. V. Miller, Trans.). Oxford: Oxford University Press.
Hegel, G. W. F. (2010). *The science of logic* (G. di Giovanni, Trans.). Cambridge: Cambridge University Press.
Heyting, A. (1956). *Intuitionism. An introduction*. Amsterdam: North-Holland Publishing Company.
Hilbert, D. (1926). Über das Unendliche. *Mathematische Annalen, 95,* 161–190.
Hilbert, D. (1930). Naturerkennen und Logik. *Die Naturwissenschaften, 18,* 959–963.
Hilbert, D. (1935). *Gesammelte Abhandlungen. Dritter Band: Analysis, Grundlagen der Mathematik, Physik, Verschiedenes. Lebensgeschichte*. Berlin: Springer.
Kant, I. (1781/1787). *Kritik der reinen Vernunft*. Riga: Johann Friedrich Hartknoch.
Kolman, V. (2009). What do Gödel theorems tell us about Hilbert's solvability thesis? In M. Peliš (Ed.), *Logica Yearbook 08* (pp. 83–94). London: College Publications.
Kolman, V. (2010). Continuum, name, and paradox. *Synthese, 175,* 351–367.
Kolman, V. (2015). Logicism as making arithmetic explicit. *Erkenntnis, 80,* 487–503.
Kolman, V. (2016a). *Zahlen*. Berlin: de Gruyter.
Kolman, V. (2016b). Hegel's bad infinity as a logical problem. *Hegel-Bulletin, 37,* 258–280.
Kolman, V. (2018). Intuition and the end of all -Isms. *Organon F, 25,* 392–409.
Kvasz, L. (2008). *Patterns of change. Linguistic innovations in the development of classical mathematics*. Basel: Birhäuser.
Lorenzen, P. (1955). *Einführung in die operative Logik und Mathematik*. Berlin: Springer.
Lorenzen, P. (1962). *Metamathematik*. Mannheim: Bibliographisches Institut.
Lorenzen, P., & Lorenz, K. (1978). *Dialogische Logik*. Darmstadt: Wissenschaftliche Buchgesellschaft.
Lorenzen, P. (1984). *Elementargeometrie*. Mannheim: Bibliographisches Institut.
Plato. Parmenides (1961). In E. Hamilton & H. Cairns (Eds.), *The collected dialogues* (F. M. Conford., Tran., pp. 920–957). Princeton, NJ: Princeton University Press.
Schütte, K. (1960). *Beweistheorie*. Berlin: Springer.
Stekeler-Weithofer, P. (1995). *Sinn-Kriterien. Die logischen Grundlagen kritischer Philosophie von Platon bis Wittgenstein*. Padeborn: Schöningh.
Stekeler-Weithofer, P. (2008). *Formen der Anschauung, Eine Philosophie der Mathematik*. Berlin: de Gruyter.
Stekeler-Weithofer, P. *Hegel's analytic pragmatism* (Unpublished). See http://www.sozphil.uni-leipzig.de/cm/philosophie/mitarbeiter/pirmin_stekeler/.
Wittgenstein, L. (1953). *Philosophical investigations*. Oxford: Blackwell.
Wittgenstein, L. (1964). *Philosophical remarks*. Oxford: Blackwell.

Chapter 8
Constructive Semantics: On the Necessity of an Appropriate Concept of Schematization

Christina Weiss

Abstract The issue of constructivity in general is by no means new, when it enters the stage of discussions concerning the foundations of mathematics at the beginning of the twentieth century, especially through L. E. J. Brouwer's supposedly radical reformation of the basic elements of mathematics. (The impression of radicalness that Brouwer obviously had on contemporaries expresses itself among others in the political names people chose for Brouwer's intuitionistic endeavour: Weyl called it "the revolution", Hilbert accordingly named it "attempted coup".) The contrary is the case: Since Kant's *Critique of Pure Reason* with its separation of the faculty of sensibility (*receptivity*) on the one side from the faculty of the intellect (*spontaneity*) on the other side, and the interconnected separation of different fields of knowledge, the *intuitive-constructive* and the *discursive* realm, (See Sect. 8.1 of the *Transcendental Aesthetic* for the introduction of the different faculties of knowledge.) a broad engagement with respect to the general form of knowledge, even and in particular of philosophy itself as a theory of knowledge, came into existence. Among others especially G. W. F. Hegel criticized the Kantian approach for its too narrow concept of constructivity. (The idealistic system-designs of Fichte and Schelling of course also comprise a criticism of Kantian thinking.) Along the main lines of Hegel's critique of Kant's limitation of constructivity to intuitive constructivity—schematic construction in pure intuition—and his closely related critique of the non-dialectical methodology of Kantian philosophy in general, we want to open up a discussion of the general form and methodology of constructive semantics as a philosophical endeavour. In this connection it turns out that Kant's idea of the necessary schematization of linguistic concepts into temporal forms of constitution, after having undergone a Hegelian revision, proves to be quite useful for clarifying the problems and challenges of and for constructive philosophy, even today.

Keywords Transcendental schematism · Homogeneity · Pure time · Universal · Singular · Dialectical schematization · Self-appearance · Pragmatics · Semantics · Dialogical constructivism · Presentational logics

---

C. Weiss (✉)
Institute of Philosophy, Technical University of Darmstadt, Darmstadt, Germany
e-mail: weiss@phil.tu-darmstadt.de

C. Weiss (ed.), *Constructive Semantics*, Logic, Epistemology, and the Unity of Science 44, https://doi.org/10.1007/978-3-030-21313-8_8

8.1 Kant's Schematism of (Pure) Concepts

In the context of examining the function of subsuming something under a certain rule, which Kant, in accordance with the Aristotelian tradition, equalizes with the faculty of judgement (see Kant 1998, B 171), Kant discusses the role of the so-called *transcendental schematism* for the application of pure concepts to empirical intuitions (see ibid., B 176). Kant assumes that subsuming something under a general rule (a concept) requires what he calls *homogeneity* of the (idea of the) concept with the (idea of the) object.[1]

For Kant, postulating the necessity of *homogeneity* between concepts and appearances—after having separated them strictly from each other—the following problem arises: How can the pure concepts of understanding—the general concepts consisting of quantitative, qualitative, relational and modal forms of judgement—which is to say, the transcendental conditions of knowledge, be applied to appearances, taking into account that pure concepts of reason cannot be sensualized at all (see ibid., B 177)?

What Kant is seeking for at this systematic point in the *Critique of Pure Reason* might be called a *method*, *medium* or *form* of integration, which in Kantian terminology is synonymous to an equalization of categorical and intuitive form. This *method*, *medium* or *form* of integration, which in Kant's eyes is a necessary condition for the application of concepts to appearances, Kant consequentially calls *transcendental schematism* (see ibid., B 177).

> Now it is clear that there must be a third thing, which must stand in homogeneity with the category on the one hand and the appearance on the other, and makes possible the application of the former to the latter. This mediating representation must be pure (without anything empirical) and yet intellectual on the one hand and sensible on the other. Such a representation is the transcendental schema (ibid., B 177).

Kant equalizes this *transcendental schema*, this "third" conceived as a mediator between general concepts (forms of judgement) and *objects-as-appearances*, which is to say, objects given in the subjective forms of intuition (space and time), with the *pure intuition of time*.

Pure intuition of time Kant introduces in the *Transcendental Aesthetic* as "nothing other than the form of inner sense, e.g., of the intuition of our self and our inner state" (ibid., B 49). This very form of inner sense, this self-intuition, Kant takes to function as "the a priori condition of all appearances in general" (ibid., A 34). That is to say, self-intuition, which Kant equalizes with time as a general form consisting of a

[1]Remark: The concept of homogeneity Kant draws on as a necessary condition for rule-application remains rather vague, especially with regard to the example Kant uses to illustrate, what he means by homogeneity. The example postulates homogeneity between the empirical concept of a plate with the geometrical concept of a circle with respect to their rounding, insofar as the (same) rounding in question would appear as part of a *thought* (in case of the empirical concept "plate") and appear in case of the geometrical circle as an *intuition* (see Kant 1998, B 176). The question what the assumed homogeneity consists in, is not answered through this example, as the example simply claims the homogeneity of thought-content and intuition-content, presupposing an already existing concept of homogeneity between intuition and thought instead of providing it.

successive change of states is *the* condition of the possibility of appearances of any shape whatsoever.

Now Kant's argument for choosing *pure time* as the mediating form between pure concepts and empirical intuitions is that time is presupposed, albeit in different manners, on the side of pure concepts as well as on the side of empirical intuitions. His argumentation goes as follows: Both sides participate in structuring, what Kant calls the manifold: The pure concepts of reason contain *synthetic unity* of the manifold as such (see ibid., B 177). Every empirical intuition is a specific structuring of the manifold (see ibid., B 177). Time as the general form of self-intuition is homogeneous with both sources of knowledge in the following sense: Time is based on a general a priori rule of succession of states. With respect to this assumed *generality of rule-following* time and pure concepts are homogeneous (see ibid., B 177). On the other hand time is by definition included in every empirical intuition. For Kant this inclusion warrants homogeneity of pure time and empirical intuition likewise.

Kant views the explication of those two "features" of pure time, *generality of form* and *being included in every specific form*—which actually is just another way of stating the generality of form—as sufficient for having shown the appropriateness of taking pure time as *the transcendental schema.*

However, the following critical questions irrefutably arise: Is the reference on the shared characteristics of generality together with the more or less tautological reference on the inclusion of pure intuition in every specific intuition sufficient to show homogeneity between pure concept and empirical intuition via time?

Is Kant's method in arriving at the homogeneity of pure concept and empirical intuition the appropriate method for understanding schematism?

The more general question in this connection of course is: What does the term "homogeneity" signify in Kant's approach? Kant actually doesn't put much effort in working out this question, but seemingly takes *homogeneity* as philosophically unproblematic. His way of dealing with it might be subsumed under the rather vague imperative: Find any concordance whatsoever between two terms! For, what actually constitutes the assumed homogeneity between pure concept and pure time? It is their generality, which is the outcome of their very definition as possibilities *sine qua non* for objective knowledge. And what constitutes the assumed homogeneity between pure time and empirical intuition? It is again the already established generality of pure time, which formally implies its being-included in every specific intuition.

What Kant should have mentioned at the least is that the relationship of homogeneity between pure concept and pure time is of a different logical form than the relationship of pure time and empirical intuition. Because for the first case, pure concept and pure time, the homogeneity consists in their shared transcendental necessity, which is equivalent to their generality. So with respect to their transcendental function pure time and pure concept are homogeneous. Bud didn't we know that before? And what does that tell us about the actual form of schematization?

Let's turn to the other relation of assumed homogeneity, the relationship between pure time and empirical intuition. We do find a different logical relationship between these two, namely that of enclosure: If pure time is by definition the form that is implied in every appearance, then it is naturally implied in empirical intuition. But what does this logical implication tell us about the actual structure of the relationship between the general form of intuition and a specific intuition?

Kant leaves those questions unanswered. What transcendental schematism actually should have shown—the form of homogeneity of concept and intuition—remains a mere formalistic postulate being satisfied with very general concepts of logical equivalence and implication.

But let's try to insist, if we can find out more about Kant's method for detecting homogeneities in the chapter on *transcendental schematism.*

He gives a hint into a different direction of thought that he unfortunately doesn't pursue further. In the paragraph just before dwelling on pure time's shared characteristic of generality and it's being-included in every empirical intuition Kant refers to the relatedness of both, category and pure time, to manifold. He writes:

> The concept of understanding contains pure synthetic unity of the manifold in general. Time, as the formal condition of the manifold of inner sense, thus of the connection of all representations, contains an a priori manifold in pure intuition (ibid., B 177).

Now instead of aiming at explicating more or less traditional logical genus-species-relationships between *pure concept*, *pure time* and *empirical concept* of the form:

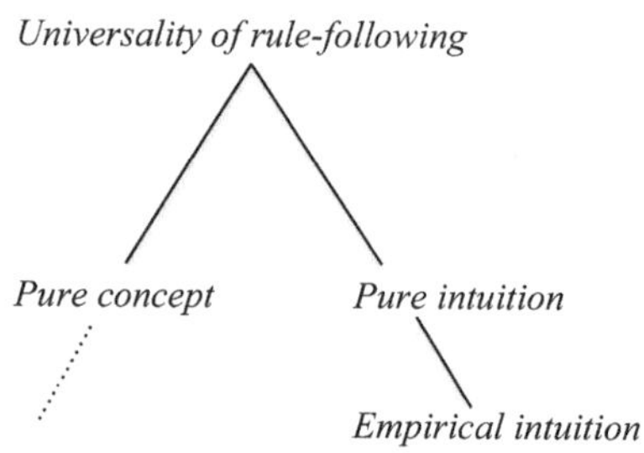

as Kant does it, one could concentrate on the very form of the relationship pure time and pure concept appear to have towards manifold. Whereas the pure concept, according to Kant, contains pure synthetic unity of the manifold as such, pure time contains a manifold a priori. What does that mean? Pure concept is the unity-constructor, disposes *form-as-unity* to everything, whereas pure time cares for the presentation of a manifold, is the *presentation-device*, which is necessary for *spatio-temporal singularization.*

Now one could say that time is related to categories as well as to appearances in the sense that time, as the a priori intuition of the manifold, is the "whereof" of the synthetic unity, which the categories constitute on the one side, the "Being-in" of every concrete appearance on the other side. Or in other words: What is *matter* in the case of the synthetic unity, that the categories constitute, is *presentation* in the case of appearances.

With this formulation, we of course already yield a more Hegelian, relational account of the form of schematization, of a possible dialectical relationship between manifold seen as *matter* and manifold seen as *presentation*. We will come back to this point further down under 2.

Unfortunately, Kant does not recognize this dialectical relationship embodied in his effort of gaining the mediating form. Therefore his concept of homogeneity, or rather his determination of time as a medium for the conceptual and the intuitive, remains formal, an external homogeneity one could say in a Hegelian jargon.

However, what renders the chapter on transcendental schematism important for the discussion on constructivity in general, although Kant's method in arriving at pure time as *the* schema as such is itself not constructive but obligated to a transcendental formalism, is:

1. The mere discovery of a necessary third, relating the linguistic forms of judgement and the forms of intuition to each other.
2. Kant's fundamental insight that the term "schema" signifies a method, namely a method of providing images, sensual figures for abstract concepts.

Ad 1. Despite of the non-constructive method of abstraction that Kant applies for gaining the transcendental schema of time, carving out the necessity of an integrating, mediating form between the linguistic and the intuitive realm of knowledge marks a cornerstone in philosophical thinking. One can convincingly argue that subsequent philosophical endeavours treat the very form of integration and relation of the intuitive and the linguistic realm as the basic problem of philosophical reasoning.

Ad 2. Together with the concentration of philosophical thinking on the explication of what Kant calls schematism, as the central problem for philosophical theory, described under 1., Kant's insight that the problem of schemata is a problem of method had a great impact on philosophical thinking as such. Especially Hegel's dialectical project turns out to combine both aspects in direction of a (re-)constructive (dialectical) theory of meaning.

8.2 The Form of Schematism

Under 1 we outlined that Kant's method, in determining transcendental schematism as the pure intuition of time, follows what might be called a logic of abstraction, that can be illustrated with the help of a traditional genus-species-tree. In the following we want to demonstrate why this method of abstraction is inadequate for what Kant is looking for in this context: the explication of a method for relating *category* and *intuition* to one another. Through the argumentation it should become clear that for an explication of this mediating form nothing less is needed than reconstructing this mediation as a process itself. The Kantian "third" then turns out to be the *dialectal process of mediation* itself, not a mere abstraction of the related aspects, *pure concept* and *intuition*.

A new perspective on *constructivity* in philosophy thus originates: *Constructivity* is not restricted to intuitive *constructivity*.[2] The contrary is the case: If philosophy wants to get insight into the relationship of *concept* and *intuition*, which is nothing

[2]Or as Kant would call it, to "schematic construction in pure intuition".

less than gaining insight into the form and process of knowledge itself, it has to account for this relational, dialectical *constructivity* in the first place.[3]

8.2.1 Kant's Schematism as a Schematism without a Schematism

It is the function of transcendental schematism in the systematics of the *Critique of Pure Reason* to restrict the use of categories to intuitions only. Pure time is meant to fulfil this task as it is, from a formal viewpoint that Kant follows in the *Critique of Pure Reason*, the mediating function between pure understanding and empirical intuition.

Although further points of criticism against transcendental formalism could be carved out we want to restrict our reconstruction of the conception of transcendental schematism and its critical reception by Hegel to merely one, but central topic:

If one accepts the problematic derivation of transcendental schematism for a moment, one should arrive at the following conclusion with respect to constructivity in philosophy:

If pure time is considered as a schematism mediating between category and empirical intuition, then pure time in the line of Kantian thought, should adopt the same function in and for transcendental philosophy that it according to Kant has together with pure space for mathematics: It should be a medium for constructing concepts. Remember that for Kant mathematical entities, that is to say, arithmetical and geometrical objects, are nothing but (schematic) constructions in pure intuition.[4]

Pure time, then, should be considered as a concept-constructor, as a construction-device for pure concepts. Because of the fact that transcendental schematism is considered as part of a transcendental theory of judgment, transcendental philosophy should accordingly possess transcendental schematism as some form of a constructive element. Construction, through transcendental schematism should enter into

[3]It might simultaneously become apparent, that all sorts of prioritizing either the intuitive or the conceptual side of knowledge together with claiming material invariances of the terms "intuition" and "concept" should be given up. See for a similar line of thought, also related to Hegel's dialectical concept of knowledge, Kolman's discussion of the either-or-dichotomy in this volume.

[4]See Kant's argumentation concerning the difference of philosophical and mathematical knowledge in the First Chapter of the *Transcendental Doctrine of Method*: "Philosophical cognition is rational cognition from concepts, mathematical cognition that from the construction of concepts" (ibid., B 741). And further down: "All of our cognition is in the end related to possible intuitions: for through these alone is an object given. Now an a priori (a non-empirical concept) either already contains a pure intuition in itself, in which case it can be constructed; or else it contains nothing but the synthesis of possible intuitions, which are not given a priori, in which case one can well judge synthetically and a priori by its means, but only discursively, in accordance with concepts, and never intuitively through the construction of the concept" (ibid., B 748). Kant here explicitly excludes the possibility for philosophy to construct concepts, although the whole paragraph on schematization actually consists of pure time's ability to construct pure concepts, that is, transform them into intuitive schemata on the transcendental level.

philosophical thought, overcoming the restriction of constructive concepts to mathematical intuition.[5]

But neither does Kant actually explicate the schematizing function of time for giving meaning to the general forms of judgement, nor—and related to the first deficiency—do his remarks on the necessity of a constructive, schematizing form in the realm of transcendental forms influence his conception of the role of constructivity in and for philosophy and the connected question of the relationship between discursive and intuitive forms of understanding.

The reverse is the case. In the *Transcendental Doctrine of Method*, which embodies a reflection on different forms of understanding, Kant separates philosophical knowledge strictly from mathematical knowledge by determining its form as exclusively discursive, whereas mathematical knowledge should consist exclusively of constructing concepts in pure intuition (see ibid., B 742–B 766). That is to say that in spite of the central role Kant attributes to transcendental schematism and despite of the fact that schematism itself is introduced as a synthesis achieved through a form of rule-following (see ibid., B 181), which actually *is* the definition of mathematical object-constitution, Kant refuses to regard his concept of schematism as a constructive philosophical element. Implied in this refusal to understand the mediating, schematizing element as a constructive element of and for philosophy, is the refusal of a "genuine" object of philosophical inquiry. Kantian transcendental philosophy, strictly speaking, is philosophy without an object of inquiry, a form of theorizing, dealing with the possibility of objects only.

> There is, to be sure, a transcendental synthesis from concepts alone, with which in turn only the philosopher can succeed, but which never concerns more than a thing in general, with regard to the conditions under which its perception could belong to possible experience (ibid., B 747).

This restriction of the domain of transcendental philosophy to the mere possibility of objects among others leaves open the relationship between actual objects and possible objects, that is the relationship between the actual content and the general form of a cognizing act.

[5]Remarkably Kant puts a lot of emphasis on the central, transcendental function of schematism: "Thus the schemata of the concepts of pure understanding are the true and sole conditions for providing them with a relation to objects, thus with **significance**, and hence the categories are in the end of none but a possible empirical use, since they merely serve to subject appearances to general rules of synthesis through grounds of an a priori necessary unity (on account of the necessary unification of all consciousness in an original apperception), and thereby to make them fit for a thoroughgoing connection in one experience" (ibid., B 185). Now, what else than a construction-process of transcendental imagination is the reason for providing significance, that is, object-relatedness for the categories?

8.2.2 Hegel's Critique of Kant's Conception of Productive Imagination

One main criticism that Hegel brings forward against Kant's conception of *transcendental imagination*, which in the *Critique of Pure Reason* is *the* faculty performing transcendental schematization, is that Kant understands the faculty of *transcendental imagination* in the systematics of *Pure Reason* only as an additional link between two independently existing faculties, that of understanding and that of intuition. Already in an early Jena essay on "Faith and Knowledge" Hegel explicates the necessity to take, what Kant calls "transcendental imagination" as a methodological point of origin, instead of introducing it as a mere connecting link between two modes of appearance of knowledge, intellect and intuition, wrongly considered in Kant's approach as two independent forces.[6]

In contrast to Kant's non-constructive positing of category and intuition with the telling consequence of a necessary but only supplementary integration through transcendental schematism, Hegel demands to conceptualize Kant's productive imagination as an originally *two-sided form*, that only secondarily appears separated into two aspects of knowledge, concept and intuition. Hegel in this context calls for nothing other than an explication of the form of *productive imagination* itself, which in Kantian eyes is the faculty responsible for schematization. In Hegel's conviction such an explication of the form of productive imagination will in turn show the original identity of concept and intuition, an identity, which actually *is* and expresses the *unity of apperception*, which according to Kant all intuitions and concepts are bound to. What Hegel demands here with regard to Kant's systematics, the Kantian type of philosophy, is nothing less than a method that actually *presents* the very form of the concepts used in philosophical theory for the description of knowledge. He criticizes Kant's formalism for a lack of demonstrative function, which in Hegel's eyes is necessary for recognizing the "real" structures of concepts and of the relationships between them. Precisely because of the fact that Kant leaves open the actual presentation of the form of his central concepts and their interconnections, but only formalistically treats their relatedness in general, the underlying structure of the process of meaning-constitution, the process of cognition itself, remains unclear.[7]

---

[6]"This original synthetic unity must be conceived, not as produced out of opposites, but as a truly necessary, absolute, original identity of opposites. [...] This shows that the Kantian forms of intuition and the forms of thought cannot be kept apart at all as the particular, isolated faculties which they are usually represented as. One and the same synthetic unity [...] is the principle of intuition and of the intellect" (Hegel 1977a, p. 70).

[7]See regarding this lack of demonstrative function that Hegel criticizes in Kant's methodology Lothar Eley's considerations on a dialectical reformulation of the Kantian trias *concept/schema/image*. With reference to Hegel's *Science of Logic* Eley argues that Hegel—overcoming the Kantian separation between intuition and concept—understands the *concept as such* as a schematic construction. Eley simultaneously considers the image-function as a diagrammatic function that actually shows the schematic constructions concepts consist in. In Eley's view, logical diagrams, for instance, do have a phenomenological, presentational function with respect to *concepts-as-schematic-constructions* (see Eley 1976, p. 29 ff.). Although we do in principle share

Returning to our general theme, the issue of *constructivity* itself, we can state the following:

Hegel in and for philosophy requires a method, which is apt to present the very form of its content, a method that *is* the very presentation of its content. The task for philosophical theory thus consists in explicating such a presentational method that (re-)constructs the identity of different aspects of knowledge—as concept and intuition—as a *presentational identity*.

This presentational, or to take Hegel's term, phenomenological method naturally cannot be imported from other scientific branches, because the explication of this method, which should be able to integrate *form and content, concept and intuition*, as a presentational unity is the very theme of philosophy itself. The specific presentational method therefore has to be gained by reconstructing the process of knowledge itself. In *The Science of Logic*, which is meant to explicate the philosophical method in general, Hegel states: "But the exposition of that which alone can be the true method of philosophical science falls within the treatment of logic itself; for method is the consciousness of the form of the inner self-movement of the content of logic" (Hegel 2010, p. 33). With respect to the question of the form of construction this also implies that mathematical concepts of constructivity, for instance those of the Kantian tradition that take construction to consist of schematizing concepts in pure, ideal intuition, cannot simply be adopted by philosophical theory, as the content of philosophical theory is a different one than the content of mathematical theory. Therefore its form, considered as a presentation of its content, has to be different.

8.2.3 Dialectical Schematization

Now what does this Hegelian critique mean for transcendental schematism, or rather for the schematizing function inside philosophical theory as such?

It should be clear that we can only offer a sketch of an answer here, as one can reasonably argue that a bigger part of Hegelian thought is an elaboration of the problems raised by Kantian philosophy, especially by his central dualisms like concept/intuition or form/content.

Therefore we want to restrict our argumentation on one, but epistemologically central impact that Hegel's criticism has induced into theorizing about schematism:

Hegel's critique of Kant's conception of the schematizing function of transcendental imagination is first and foremost a critique of Kant's method of reasoning. In the context of explaining Kant's choice of pure time as adopting the mediator-function between category and empirical intuition, we already glimpsed at a different possibility of relating category and empirical intuition, when we argued that the form

the conviction of a dialectical relationship between diagrams and concepts we do want to remark that this very demonstrative function of diagrams with respect to concepts needs to be elaborated further than Eley does it at this point. See related to the question of diagrammatic constructions and concepts (Weiss 2018).

of time, from the "perspective" of the general forms of judgement is the "whereof" or matter of their synthetic unity, whereas from the perspective of empirical intuition, time is the form, in which every specific intuition presents itself (see p. 176 in this volume).

We already mentioned that Kant introduces the medium in a different way, via abstraction, which in its form as well as in its result remains quite unsatisfying.

Now Hegel's own methodology in describing the relationship of what Kant calls concept and what he calls intuition, or rather, in his own theory of the functions and forms of knowledge, first of all performs a fundamental change in perspective on the problem addressed by Kant.

Instead of asking which of the alleged different faculties of cognition should be able to mediate between transcendental concepts and empirical intuitions, Hegel takes the form of knowledge to be constitutively of an (self-)appearing, self-revealing shape.[8] Knowledge, in Hegel's view is not considered along the lines of a toolbox, with which the cognizing subject forms something else, the manifold, conceived as something foreign, fundamentally different from the cognizing subject. In contrast to this view Hegel conceptualizes and describes knowledge as an integrative process of self-reflection, self-revelation, in which different levels of meaning-constitution do not describe different faculties of human understanding, but different grades and stations of self-appearance.[9]

In sharp difference to the toolbox-conception of cognition different grades and stations of the self-appearance of knowledge are integrated, operationally and systemically related to another from the beginning through the very process of self-revealing, which is the and the only place of their being.

We cannot go to deep into Hegel's conception of this self-revealing process, the function and form of *negation* and *sublation*, the specific form of self-reference embodied in Hegel's conception of the self-differentiation of non-differentiated, self-identical forms, because that would require a reconstruction of his central philosophizing, his philosophical method as such.[10] Therefore we only want to sketch, what Hegel's simultaneously self-reflexive and phenomenological turn implies for the process of schematization.

---

[8]See concerning Hegel's conception of philosophical form as self-construction (Hegel 2010). "This spiritual movement, which in its simplicity gives itself its determinateness, and in this determinateness gives itself its self-equality – this movement, which is thus the immanent development of the concept, is the absolute method of the concept, the absolute method of cognition and at the same time the immanent soul of the content. – On this self-constructing path alone, I say, is philosophy capable of being objective, demonstrative science" (ibid., p. 10). To avoid the typical misunderstanding in relation to Hegel's terminology it is important to stress that the term "the absolute" has a strict methodological meaning in Hegel's logic. It signifies a self-conditioning and self-recognizing procedure.

[9]See for the shape this development has from the viewpoint of what Hegel calls "natural consciousness" (Hegel 1977b). "[…] An exposition of how knowledge makes its appearance will here be undertaken" (ibid., p. 49).

[10]Readers profoundly interested in the Hegelian method of thinking should work their way through Hegel's main works like the *Phenomenology of Spirit* or *The Science of Logic*.

8.2.3.1 Schematization as Self-schematization

Hegel's uncovering of the unity of different aspects of knowledge like *intuition* and *concept* necessarily leads to a change of direction in philosophical methodology. Hegel takes the insight into the unity of different (re)presentational forms, in which knowledge appears in the process of its manifestation, as a methodological guideline for the reconstruction of the self-differentiation of knowledge. In this connection the unity of different forms of (re)presentation of knowledge becomes a reference point in Hegel's line of thought. As an example of this organisation of philosophical method around the unity of representative forms one can take Hegel's *Phenomenology of Spirit* with its integrative movement between what Hegel calls *sense-certainty* and what he calls *absolute knowledge* (see Hegel 1977b). In a way the movement explicated by Hegel consists in checking the appropriateness of each form of (re)presentation with regard to the meaning it intends to express.[11] The function of the dialectical movement in this context it is to reconstruct the process of knowledge itself as a process of unification. So unity does not remain a mere postulate, as in Kantian philosophy, but is itself being reconstructed as the key impetus of meaning-constitution. What in Kantian thinking appears as a mere repair work for an originally separative thinking, in Hegel's systematics advances to the key and core methodological device for philosophical theory: *(self-)schematization of knowledge*.

It is of central importance to emphasize that this change of direction does not only consist in a change of the order of argumentation—*from the unity of knowledge to its different parts* instead of *from the different parts to the unity*. In fact Hegel's explication of the unity of knowledge comprises a fundamental change with respect to the conception of knowledge, including philosophical theory. For the unity of knowl-

[11]It is important to emphasize that although pragmaticist Robert Brandom in *Wiedererinnerter Idealismus* (Brandom 2015) seems to follow a similar path of understanding Hegel's dialectical method, our understanding of what we call "appropriateness check" differs profoundly from Brandom's inferentialist interpretation. For Brandom understands Hegel's dialectics more or less as a variant of the positivistic method of trial and error. For him the Kantian *unity of apperception* consists in the consistency of a (deductive) logical system. In other words: If two assumptions are from the viewpoint of classical logics incompatible with each other, something must be undertaken to get rid of this incompatibility. If somebody is convinced that something *is* a dog and likewise is convinced that the same thing *is* a cat, either one of both assumptions is false or both are false, but they cannot be true together, because, as Brandom calls it, *being a cat* materially excludes *being a dog* (see Brandom 2015, Chap. 5 and for the concept of *material imcompatibility* Chap. 4). So according to Brandom, what pushes the process of cognition from mere appearance towards reality (Wirklichkeit) is a sort of implicit consistency test that has a normative function for an actor's beliefs. This trivializing understanding of *negation*, of *being in itself* and *being for itself*, of *appearance and reality* and other Hegelian terms doesn't capture Hegel's philosophical method. What might instead be understood by the technical term "appropriateness check" is Hegel's built in reflection on the appropriateness of a certain form of presentation for a content, that is the appropriateness of and for a self-representation of a content, which is nothing less than the self-construction of form. The guiding question concerning the appropriateness of presentation in Hegel's method is: Does the respective presentation reflect its content as such? Is it as a form of presentation a result of the self-presentation of the content? "Content" in Hegel's view of course cannot be equalized with (empirical) proposition. "Content" in general means cognition, at particular stages, for example in *Phenomenology of Spirit* it signifies different stations of the self-presentational process of cognition.

edge consists in the unity of self-presentation, self-description of knowledge. Only in considering knowledge as a self-presentational form, and in actually (re)constructing this self-presentation does the unity, in the form of a relational unity between the different modi of presentation, exist.

This phenomenological characteristics of knowledge, the constitutive role of (self-)appearance in and for the process of knowledge transforms the meaning and function of the terms *intuition* and *concept* fundamentally. In Hegel's systematics *intuition* and *concept* become aspects in the self-appearance and self-description of knowledge. Through their embeddedness in the cognizing operation their (ontological) status changes profoundly. Whereas Kant understands *intuition* and *concept* merely psychologistically as mental faculties, whose function consists in the conversion of the manifold into knowledge, in Hegel's conception they become *indicating forms* of and for a self-conscious operation.[12]

This transformation of the Kantian *faculties* into internally related *signs* of a self-reflexive cognizing operation simultaneously renders possible what might be called a flexibilization of the meaning of the different aspects of cognition.[13]

The term "flexibilization of meaning" in this context is used to stress the difference to the Kantian approach with its semantically fixed contents of category (forms of judgement) and intuition (Newtonian space and time). In contrast to that the Hegelian shapes of cognition, which are conceived of as signifying shapes of a reflexive process of cognition, are capable of diverse contextualizations. This flexibility preserves Hegel's approach from the Kantian destiny of becoming implausible with respect to the obvious historicity of his a priori conditions of knowledge.

8.3 Presentational Logic and Modern Constructivism

Although Kant indeed as one of the central ideas of his theoretical philosophy carries out a transcendental-philosophical re-interpretation of *concept* and *intuition* as *conceptual and intuitive conditions* of the possibility of knowledge, that is an epistemological re-functionalization of traditional logic on the one side, Newtonian conceptions of space and time on the other side, he adopts the traditional logical form and the physicist concepts of space and time without significant changes. In contrast to this Hegel demands a constitutive binding of any epistemological function to the form of its philosophical (re)construction. For philosophical theory this sets higher requirements than merely postulating "necessary conditions" for possible knowledge. It necessitates any shape and function of knowledge to emerge out of the process of self-description, as which knowledge is conceptualized in the Hegelian approach. One central demand of Hegel's theorizing in this connection consists in his insisting

[12]That is, the different aspects of knowledge are distinct, not separated, with respect to their indicating and presentational function. This thought is elaborated in Peirce's semiotic epistemology.

[13]See Kolman in this volume, who develops a similar Hegelian argument against the ahistoricity of intuitionism in the shape of Brouwer.

on what might be called *a philosophical form of theory*, that should be apt to mirror the claimed genetic connection between different aspects of knowledge.

Under 2 we outlined that the Hegelian *conversion of the mode of thinking* turns into a requirement of a theory of self-schematizing knowledge.

Now, as a matter of course, the question arises, what significance such an enlarged, namely dialectical conception of construction might have for theorizing in and about constructive semantics in present surroundings. It should be clear that because of the limited scope of an article we can only outline possible implications for constructive semantics.

8.3.1 The General Form of Constructive Semantics

Tying in with the above discussion we want to draw our attention on the general form of a constructive theory of meaning-constitution. That is to say that we don't want to address the more specific logical aspects of Brouwerian or dialogical interpretations of constructivism in this place, but focus on the form of (re)presentation of constructive semantics as a general philosophical theory.

8.3.1.1 Dialogical Constructivism

Recent developments in constructive philosophy, especially those brought forward by Kuno Lorenz in his works on *Dialogical Constructivism* (see Lorenz 2009a) articulate the insight that for a general constructive epistemology the explication of a general form of pragmatics is necessary. From the perspective of general epistemology, prior to already formalized logical systems and their possible constructive interpretations, the question of *doing* and *understanding*, *action* and *conception*, must be given a general constructive account.[14] That is to say, instead of solely offering constructive interpretations of specific elements of formal systems of logic, for instance of the junctors of propositional logic, as a general constructivist philosophy, dialogical constructivism seeks to constructively explain the genesis of *object-world and language-system*, without restricting the universe of discourse to mathematical or formal objects. In this regard Kuno Lorenz's thoughts towards a general dialogical philosophy follow Peirce's pragmatic-semiotic epistemology, whose advantage over the traditional constructivism of the *Erlangen-school* according to Lorenz lies precisely in its generality concerning the objects of discourse (see Lorenz 2008, p. 22).

One main concern Lorenz is pursuing in this context is a pragmatic-semiotic foundation of key epistemological concepts like *object*, *concept*, *meaning* and *sign*

[14] See concerning the foundations of the so-called Erlanger Konstruktivismus (Kamlah and Lorenzen 1987; Mittelstraß 2008; Lorenz and Mittelstraß 2011).

together with the related object-logic of *universal*, *singular* and *particular* judgements (see Lorenz 2009b).

Methodologically similar to the *Erlangen-approach* Lorenz starts his pragmatic reconstruction of object-world and language-system—shortly *meaning* and *sign*—with non-dispensable elements of a dialogue. Based on these non-dispensable elements like *activity/passivity* or *doing/understanding* through the built-in possibility of self-application he then continues to introduce more abstract epistemological concepts like *sign* and *object* (see Lorenz 2009b).[15]

Functionally equivalent to the *elementary predication* that the *Logical Propaedeutic* uses for a signification of a not yet formal, indispensible form of predication, Lorenz calls these indispensible elements of a dialogue as a whole "elementary learning and teaching situation" (see Lorenz 2009b).

The elementary learning and teaching situation consists of dialogical invariants of meaning-constitution like *active side of action/passive side of understanding, schema/actualization, instantiation (exemplification)/adduction (generalization)* (see Lorenz 2009b, 31 ff.). The form of these invariants is that of binary complements, complementary roles in an idealized social interaction. Now the crux of these complementary roles in the idealized social game is that they mirror proto-quantificational aspects of judgements on the level of idealized social roles. Lorenz connects semantic distinctions of *general concept* and *singular case* with dialogical roles, or better, he tries to reconstruct semantic distinctions starting with dialogical roles.

The execution of an action—Lorenz takes the example of swimming—he equalizes with an *instantiation of a schema*. An actor's swimming singularizes the schema swimming: "What I am doing is an example of swimming." On the other hand the passive reception, observation of a singularizing operation schematizes, that is *generalizes the singular operation*: "What she is doing there is swimming." (see Lorenz 2009b, p. 32).

It is important to stress, that *singularization* and *schematization* are considered as two aspects of one action, not two different sorts of action. What aspect comes into the foreground of observation is completely dependent on the viewpoint under which it is regarded: For the dialogical role of teaching one needs to *singularize a schema*, for the dialogical role of learning one needs to *understand an action* (see Lorenz ibid., p. 30 ff.). That is to say that any action in the *elementary learning and teaching situation* literally contains both sides, the active instantiating side and the passive generalizing side. Lorenz himself emphasizes that on the level of *elementary learning and teaching* the terms "to actualize a schema" and "to schematize an action" must be taken to be strictly synonymous, formulated from different viewpoints only (see

[15]Recall that in *Logical Propaedeutic* it is the so-called *elementary predication* that serves to function as a non-dispensible element for reconstructing the language of ordinary formal logic out of everyday distinguishing (see Kamlah and Lorenzen 1987, Chap. 1). Kant's transcendental heritage is thus definitely recognizable in the methodological principle of Erlangen- and dialogical constructivism.

ibid., p. 32). "Jede Prähandlung hat eine pragmatische und eine semiotische Seite" (Lorenz ibid., p. 33).[16]

So in the elementary teaching and learning situation Lorenz considers, or better is forced to consider a form, in which and for which pragmatic and semiotic aspects are not distinguished yet, but which renders it possible, to differentiate further into what Lorenz calls *action-schema* and *object-schema*, that is pragmatic and semantic aspects of meaning-constitution (see Lorenz ibid., p. 33 ff.; see also Scherer 1984, p. 58 ff.).

We don't want to reconstruct the whole differentiation-process, which is meant to lead into a dialogical reconstruction of predication at this place. Because of this article's interest in discussing the general form of constructivity starting with Kant's schematization-problem, we want to concentrate our further considerations on this pragmatic-semiotic form that Lorenz exposes in the *elementary learning and teaching situation* as a starting point for his reconstruction.

Our major point of interest in this context is the conception of the *pragmatic-semiotic unity* and—related to the first—the operations carried out with and on the *pragmatic-semiotic unity* for differentiation purposes.

Lorenz actually doesn't offer an elaborated theorizing on the form of the *pragmatic-semiotic unity*, but adheres to Peirce's pragmatic conception of meaning and sign-constitution, in which the relative position in an iterative process of sign/meaning-constitution decides, what function an element has in and for an action (see also Scherer ibid., Chap. 2.3).

What can be stated at the least concerning the *form of form* Lorenz introduces is that there is a dialectical relationship embodied in the concept of form. If a (pre-)action as a form embodies both, a pragmatic and a semiotic side, and those two sides only exist in a reciprocal constitutional dependence to and from each other, this form deserves to be called a *dialectical form*, or better a *dialectics of form*.

Under 2.3 we outlined basic characteristics of such a dialectical concept of form with respect to the Kantian problem of transcendental schematization. We stressed that Hegel's dialectical unity of concept and intuition via dialectical schematization profoundly differs from the conception of unity Kant has in mind, when he introduces time as *the* transcendental mediating schema. Hegel's concept of unity is a relational, presentational one, whereas Kant's concept of unity is abstract and psychologistic. One important point in this context is that the process of dialectical schematization cannot be modelled with the help of genus-species-trees. Now the same obviously holds for Lorenz's concept of form as *action/sign*. The action/sign-complementarity as a complementarity cannot be captured by genus-species-trees, because one cannot state that what functions as a singular action does not function as a universalizing sign. The contrary is the case: Only that, which can be singularized through instantiating actions, can have the function of a sign, vice versa.

[16]"Every pre-action contains a pragmatic and a semiotic side" (translation by C. Weiss). Lorenz calls the action a pre-action, because at this point subjective and objective side of action have not been differentiated yet.

But Lorenz doesn't fully recognize the dialectical logic sitting in the back of his considerations. In any case he underestimates the implications of his dialectical concept of action and sign. Instead of noticing the obvious resemblance to Hegel's concept of meaning-constitution and although Lorenz himself calls the form of pre-action a janus-faced one (see Lorenz ibid., p. 32) Lorenz proceeds in his process of reconstruction without elaborating on this prominent similarity to Hegel's considerations.

8.3.1.2 The "Principle" of Self-similarity

Now one central if not the most important feature of the *elementary learning and teaching situation* consists in its built-in capacity of *self-application*, which Lorenz calls "following the principle of self-similarity" (transl. C. Weiss, see Lorenz 2009c, p. 106).[17] In spite of the fact that he actually connects, what he calls "a repetition of method" [(Wiederholung des Verfahrens) Lorenz 2009b, p. 32] to the janus-faced structure of schema-instantiation and schema-generalization, he doesn't work out the dialectical foundation of this structure. What he wants to carve out is that the complementary roles of instantiation/generalization are being further differentiated by exactly applying the procedure of instantiation/generalization to each side of the difference-identity. "Erst dadurch, daß in einem Iterationsprozeß durch Selbstanwendung, was als Befolgen eines Prinzips der 'Selbstähnlichkeit' verstanden werden kann, die beiden dialogischen Perspektiven in eigenständige Handlungen überführt werden, die dann ihrerseits beiden Perspektiven unterworfen sind, lassen sich nacheinander die für den Weg zur Sprache erforderlichen Verfeinerungen in die Modellbildung einbringen" (Lorenz 2009c, p. 105 f.).[18]

We do generally agree that differentiation is a form of self-application.[19] Nevertheless we want to go a step further and ask explicitly why this complementary structure of action/sign renders self-application possible if not necessary.

So let's look closer at the action/sign-distinction. As we already pointed out Lorenz stresses its obvious and indispensible complementarity. In his words: *Actualizations* can only be understood referring to their *schema*, and a *schema* is present only in its *actualizations* (see Lorenz 2009b, p. 32). One might say that both, singularizing action and generalizing understanding are inherently bound to a predicative form of the shape *something is something*:

1. "*This* (*what I do*) is (an example of) swimming." = *singularizing action*
2. "What you do over there is *swimming*." = *generalizing understanding*

---

[17]"[...] Was als Befolgen eines Prinzips der >Selbstähnlichkeit< verstanden werden kann" (Lorenz ibid.).

[18]"Only through converting both dialogical perspectives into two independent actions, which on their part become subjected to the two perspectives, in an iterative process through self-application – which can be understood as obeying the principle of ‚self-similarity' – the necessary refinements for reconstructing the path to language can be included into the modelling" (translation C. Weiss).

[19]See for a formal concept of self-referential schemata (Weiss 2006, Chap. 3).

So *singularizing action* in its very structure consists of the action/sign-distinction as well as *generalizing understanding* consists of the action/sign-distinction. One might call this a form of *nested distinctions*, of *built-in self-referentiality* or *re-entry of distinctions into distinctions*, if one wants to carve out similarities of this concept to general aspects of formalizing distinctions.[20] But we don't want to enter into further considerations about the self-referentiality of form at this point. The only thing that we do want to emphasize with respect to our overall theme, *constructivity*, is that the reason for the possible self-application of the action/sign-distinction that Lorenz pronounces, lies in its already *self-applicatedness antecedent to every concrete application*. In other words: Self-application belongs to the form of action/sign, action/sign *is* self-application in its form.

How can we understand this structural circumstance in relation to the question of meaning-constitution? If self-application literally *is* the form the social game called *elementary learning and teaching situation* consists of, then the action/sign-form must be understood as a *play-rule* in the following sense:

> Every concept that you use must be instantiable, every instantiation that you observe must be generalizable!

As we already mentioned this circumstance might be interpreted as a built-in restriction to predicativity. But in addition, albeit in a peculiar way related to built-in predicativity, this play-rule consists in the demand *to schematize the form-as-unity into a differential unity* of a self-revealing shape, that is to execute, what the social game consists of.

8.3.1.3 Presentational Logic

The similarities to Hegel's conception of dialectical schematization are obvious. Simultaneously, through Lorenz's commitment to specific invariances of *action* and *sign*, *instantiation* and *adduction* of a schema Hegel's general concepts of the self-revelation of knowledge, of appearing knowledge, that is his presentational concept of knowledge, might be claimed to receive a concrete, schematized model in dialogical constructivism, as the dialectics of *individualizing action/generalizing understanding* opens up an understanding for what the presentational (formerly intuitive) aspect and what the conceptual aspect of knowledge regarded from the viewpoint of modern pragmatics might look like. The presentational side as the singularizing action might be understood using the following imperatives:

Show your ability to bind your actions to concepts by letting your actions become appearances of concepts! That is, let all your actions be actions of something!

On the other hand the conceptualizing side might be understood using the following imperatives:

[20] See for a formalization of the process of doing and signifying (Spencer-Brown 1972); see for a dialectical reading of Spencer-Brown's *Laws of Form* (Weiss 2006, 2018).

Relate every action you observe to a general concept, which is of course synonymous to conceptualize performances as actions! Don't observe something without relating it to something!

As a matter of fact both sides, the pragmatic and the semantic side, contain imperatives sharing the same content, a circumstance which enables us to draw the following Hegelian conclusion: What function an element has in and for the process of meaning-constitution depends on the pragmatic/semantic-context, that is, it depends on the (dialogical) question that the particular element is an answer for.[21]

8.3.1.4 Self-application of Semantic-Pragmatic Unity

In addition to carving out the dialectics implied in Lorenz's concept of form the general question of the relationship between semantic and pragmatic concepts for a theory of meaning-constitution must be addressed. If the action/sign-unity is indeed considered to be indispensible for the reconstruction of object-world and language-system then this result should have an impact on the form of philosophical semantics itself. We have shown elsewhere that it is principally impossible to found meaning-theory on pragmatic knowledge (knowing how) only, as every pragmatic competence necessarily must be accompanied by a reflexive (knowing that), semantic competence to count as a competence of and for something at all (see Weiss 2018).

We cannot go into too much detail of the dialectical relationship between semantic and pragmatic aspects of theorizing about the general form of meaning-constitution at this point. What we do want to mention here is that the insight into the indispensability of the pragmatic/semantic-distinction should lead to a fundamental reflection of the role of phenomenology for present theories of meaning-constitution.

Lorenz himself repeatedly mentions the important role of *phenomenological reduction*, that is the necessarily reductive acquisition of that experience, which has been constructed dialogically.

> Die dank der Handlungs- und Sprachkompetenz bereits mitgebrachte Erfahrung zu verstehen erfordert nämlich eine phänomenologische Reduktion durch Abblenden der vorliegenden Gliederungen vermöge eines Eingriffs und damit einer ausdrücklich hervorgehobenen Gliederung dieser Erfahrung mithilfe zunächst einfacher Handlungen und Sprachhandlungen als Mittel (Lorenz 2009b, 26 f.).[22]

Although using the term "phenomenological reduction" extensively Lorenz doesn't understand it in a Husserlian way, that is, as a method of illustrating eidetic, invari-

[21] From this viewpoint the most important lesson of *elementary learning and teaching situation* is the rule to regard, do, perceive *something as something*. This is a far more important rule than learning so-called empirical concepts, because it instantiates social form and meaning-form, or better social form as meaning-form, meaning-form as social form.

[22] "The experience of understanding, that is available because of action-competence and language-competence, necessitates a phenomenological reduction through an intervening screening off of existent classifications, and thereby an explicit accentuation of this experience. This phenomenological reduction is executed with initially simple actions and actions of speaking" (translation C. Weiss).

ant structures of transcendental consciousness. In contrast to that Lorenz obviously organizes his understanding and use of *phenomenological reduction* along the lines of Peirce's phenomenological introduction of the categories of *firstness, secondness and thirdness* in which, according to Scherer, phenomenology, or better the phenomenological categories serve to provide different approach-forms to objects (see Scherer 1984, p. 41). At least Lorenz understands *phenomenological reduction* as a form of a schematizing understanding of a dialogical action (see Lorenz 2009b). That is, *phenomenological reduction* in Lorenz's account takes the role of an *operational invariant* rather than that of a method of carving out invariants of transcendental consciousness. *Phenomenological Reduction* is meant to symbolize a schematizing action, which is necessary in the pragmatic reconstruction of an object-world.[23]

Although we do in general agree with such an attempt of an operational re-interpretation of phenomenological terms, we do want to mention that the role and form of phenomenology in and for dialogical constructivism remains quite unclear: For instance, where do central semantic distinctions contained in the operation of phenomenological reduction like *dimming out* certain aspects *and highlighting others* come from? By which method do we know that we do have a *fore-* and a *background* of experience? Are these semantic elements taken out of everyday speech? And if so, what is their epistemological justification?

In any case Lorenz doesn't put enough effort in clarifying those open questions. We do want to mention that obviously no monolithic account, neither one that privileges the semantic side, nor one that privileges the pragmatic side of cognition will succeed in this direction, but only a relational, dialectical account.

Phenomenological logician Lothar Eley for instance has convincingly argued in favour of the constitutive role of phenomenological terms for gaining and understanding the term "construction". To summarize his complex position at this point:

1. Without the semantic distinction of *meaning-intention and meaning-fulfilment* it is impossible to carve out a concept of construction, at least if this is meant to be of any epistemological value.
2. Without a constructive interpretation of *meaning-intention* and *meaning-fulfilment* in the form of offering a presentational side for the linguistic concepts in use, the relationship of *meaning-intention* and *meaning-fulfilment*—language and object—remains a mystery (see Eley 1985 and the *introduction* of this volume).

Integrating dialectical phenomenology with dialogical constructivism thus appears to be a reasonable step towards *Constructive Phenomenology*.

---

[23] Thanks to the reviewer of this article, we came across a very interesting elaboration oft the epistemic role of intuition in the lines of dialogical constructivism by Heinzmann (2013). Unfortunately we couldn't implement Heinzmann's results here. But we do want to point to the fact that dialogical constructivism actually works on developing a general epistemological account of the relationship between conceptual and intuitive elements of meaning-constitution.

8.4 Concluding Remark

It was the overall goal of this article's argumentation to show, that the question of constructivity in philosophy is far from being restricted to offering constructive proofs, constructive introductions for logical or mathematical elements. Starting with Kant's conception of *transcendental schematism* and Hegel's critical revision of it, we demonstrated that the question of constructivity is deeply connected to the question of the *form of philosophical theory* in general. With respect to Hegel's position on the inseparability of intuitive and conceptual aspects of knowledge we outlined, what could be called a concept of dialectical schematization of semantic and pragmatic aspects of meaning-constitution for the object-level of philosophical theory and simultaneously for the level of philosophical theory itself. Referring to Kuno Lorenz's concept of dialogical constructivism, we firstly expatiated on fundamental similarities to Hegel's dialectical conception of knowledge. Secondly we argued that Lorenz underestimates the theoretical consequences of the dialectical relationship of his basic difference-identities like *action/concept*. Especially with respect to the relationship of phenomenological and pragmatic elements inside the theory of meaning-constitution we emphasized the necessity of a revision that strengthens the role of phenomenological elements as part of what we call *presentational logic*.

References

Brandom, R. (2015). *Wiedererinnerter Idealismus*. Berlin: Suhrkamp.

Eley, L. (1976). *Hegels Wissenschaft der Logik*. München: Wilhelm Fink.

Eley, L. (1985). *Philosophie der Logik*. Darmstadt: Wissenschaftliche Buchgesellschaft.

Hegel, G. W. F. (1977a). *Faith and knowledge* (W. Cerf & H. S. Harris, Trans.). New York: State University of New York Press.

Hegel, G. W. F. (1977b). *Phenomenology of spirit* (A. V. Miller, Trans.). Oxford: Oxford University Press.

Hegel, G. W. F. (2010). *The science of logic* (G. di Giovanni, Trans. and Ed.). New York: Cambridge University Press.

Heinzmann, G. (2013). *L'intuition épistémique. Une approche pragmatique du contexte de compréhension et de justification en mathématiques et en philosophie*. Paris: Vrin.

Kamlah, W., & Lorenzen, P. (1987). *Logische Propädeutik. Vorschule des vernünftigen Redens*. Mannheim: Bibliographisches Institut.

Kant, I. (1998). *Critique of pure reason* (P. Guyer & A. W. Wood, Trans. and Ed.). Cambridge: Cambridge University Press.

Lorenz, K. (2008). Das Vorgefundene und das Hervorgebrachte. Zum Hintergrund der >Erlanger Schule< des Konstruktivismus. In J. Mittelstraß (Ed.), *Der Konstruktivismus in der Philosophie im Ausgang von Wilhelm Kamlah und Paul Lorenzen* (pp. 19–31). Paderborn: Mentis.

Lorenz, K. (2009a). *Dialogischer Konstruktivismus*. Berlin: Walter de Gruyter.

Lorenz, K. (2009b). *Artikulation und Prädikation*. In: K. Lorenz (Ed.), pp. 24–71.

Lorenz, K. (2009c). *Grammatik zwischen Psychologie und Logik*. In: K. Lorenz (Ed.), pp. 94–117.

Lorenz, K., & Mittelstraß, J. (2011). *Philosophische Variationen*. Berlin: Walter de Gruyter.

Scherer, B. M. (1984). *Prolegomena zu einer einheitlichen Zeichentheorie*. Tübingen: Stauffenburg-Verlag.

Spencer-Brown, G. (1972). *Laws of Form*. New York: The Julian Press.
Weiss, C. (2006). *Form und In-formation. Zur Logik selbstreferentieller Strukturgenese*. Würzburg: Königshausen & Neumann.
Weiss, C. (2018). Towards a phenomenology of schematization. *Cybernetics and Human Knowing, 24,* 245–260.

Kolibri 360 GmbH
Ferdinand-Jühlke-Straße 7
99095 Erfurt, Deutschland
produktsicherheit@kolibri360.de